National Center for Construction Education and Research

Plumbing Level Three

Prentice
Hall

Upper Saddle River, New Jersey
Columbus, Ohio

contren™
Learning Series

National Center for Construction Education and Research

President: Dan Bennet
Vice President of Training Operations and Program Development: Don Whyte
Director of Curriculum Revision and Development: Daniele Dixon
Plumbing Project Manager: Deborah Padgett
Production Manager: Debie Ness
Copy Editor: Rebecca Hassell
Desktop Publisher: Jessica Martin

NCCER would like to acknowledge the contract service providers for this curricula: EEI Communications, Alexandria, Virginia; and Document Technology Resources, Manassas, VA.

This information is general in nature and intended for training purposes only. Actual performance of activities described in this manual requires compliance with all applicable operating, service, maintenance, and safety procedures under the direction of qualified personnel. References in this manual to patented or proprietary devices do not constitute a recommendation of their use.

10 9 8 7 6 5 4 3
ISBN 0-13-061606-0

Preface

This volume was developed by the National Center for Construction Education and Research (NCCER) in response to the training needs of the construction and maintenance industries. It is one of many in NCCER's *Contren™ Learning Series*. The series, covering close to 40 craft areas and including all major construction skills, was developed over a period of years by industry and education specialists. Sixteen of the largest construction and maintenance firms in the United States committed financial and human resources to the teams that wrote the curricula and planned the nationally accredited training process. These materials are industry-proven and consist of competency-based textbooks and instructor's guides.

NCCER is a non-profit educational entity affiliated with the University of Florida and supported by the following industry and craft associations:

PARTNERING ASSOCIATIONS

- American Fire Sprinkler Association
- American Petroleum Institute
- American Society for Training and Development
- American Welding Society
- Associated Builders and Contractors, Inc.
- Associated General Contractors of America
- Association for Career and Technical Education
- Carolinas AGC, Inc.
- Carolinas Electrical Contractors Association
- Citizens Democracy Corps
- Construction Industry Institute
- Construction Users Roundtable
- Design-Build Institute of America
- Merit Contractors Association of Canada
- Metal Building Manufacturers Association
- National Association of Minority Contractors
- National Association of State Supervisors for Trade and Industrial Education
- National Association of Women in Construction
- National Insulation Association
- National Ready Mixed Concrete Association
- National Utility Contractors Association
- National Vocational Technical Honor Society
- North American Crane Bureau
- Painting and Decorating Contractors of America
- Portland Cement Association
- SkillsUSA-VICA
- Steel Erectors Association of America
- Texas Gulf Coast Chapter ABC
- U.S. Army Corps of Engineers
- University of Florida
- Women Construction Owners and Executives, USA

Some of the features of NCCER's *Contren™ Learning Series* are:

- A proven record of success over many years of use by industry companies.
- National standardization providing portability of learned job skills and educational credits that will be of tremendous value to trainees.
- Recognition: upon successful completion of training with an accredited sponsor, trainees receive an industry-recognized certificate and transcript from NCCER.
- Compliance with Apprenticeship, Training, Employer, and Labor Services (ATELS) requirements (formerly BAT) for related classroom training (CFR 29:29).
- Well-illustrated, up-to-date, and practical information.

FEATURES OF THIS BOOK

Capitalizing on a well-received campaign to redesign our textbooks, NCCER is publishing select textbooks in a two-column format. *Plumbing Level Three* incorporates the design and layout of our full-color books along with special pedagogical features. The features augment the technical material to maintain the trainees' interest and foster a deeper appreciation of the trade.

Did You Know? explains fun facts and interesting tidbits about the plumbing trade from historical to modern times.

On the Level provides helpful hints for those entering the field by presenting tricks of the trade from plumbers in a variety of disciplines.

We're excited to be able to offer you these improvements and hope they lead to a more rewarding learning experience.

As always, your feedback is welcome! Please let us know how we are doing by visiting NCCER at www.nccer.org or e-mail us at info@nccer.org.

Acknowledgments

This curriculum was revised as a result of the farsightedness and leadership of the following sponsors:

Camden County High School   Jones County High School
Encompass Mechanical Services SE   Plumb Works
JF Ingram State Technical   TD Industries

This curriculum would not exist were it not for the dedication and unselfish energy of those volunteers who served on the Authoring Team. A sincere thanks is extended to:

Jonathan Byrd   Jonathan Liston
Ed Cooper   Charles Owenby
Carlos Jones, Sr.   Ken Swain

Contents

Applied Math

COURSE MAP

This course map shows all of the modules in the third level of the Plumbing curriculum. The suggested training order begins at the bottom and proceeds up. Skill levels increase as you advance on the course map. The local Training Program Sponsor may adjust the training order.

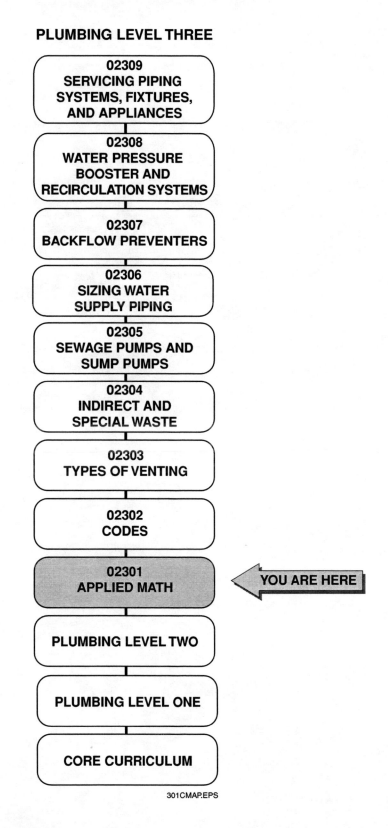

PLUMBING LEVEL THREE

02309
SERVICING PIPING SYSTEMS, FIXTURES, AND APPLIANCES

02308
WATER PRESSURE BOOSTER AND RECIRCULATION SYSTEMS

02307
BACKFLOW PREVENTERS

02306
SIZING WATER SUPPLY PIPING

02305
SEWAGE PUMPS AND SUMP PUMPS

02304
INDIRECT AND SPECIAL WASTE

02303
TYPES OF VENTING

02302
CODES

02301
APPLIED MATH   ◄ **YOU ARE HERE**

PLUMBING LEVEL TWO

PLUMBING LEVEL ONE

CORE CURRICULUM

301CMAP.EPS

MODULE 02301 CONTENTS

Figures

Tables

Applied Math

Objectives

When you have completed this module, you will be able to do the following:

1. Identify the weights and measures used in the English and metric systems.
2. Demonstrate an understanding of the concepts of area and volume.
3. Demonstrate an understanding of the practical applications of area and volume calculations.
4. Demonstrate an understanding of the concepts of temperature and pressure and how they apply to plumbing installations.
5. Explain the functions and applications of six simple machines.

Prerequisites

Before you begin this module, it is recommended that you successfully complete the following: Core Curriculum; Plumbing Level One; Plumbing Level Two.

Required Trainee Materials

1. Calculator
2. Pencil and paper

1.0.0 ◆ INTRODUCTION

Plumbers use math every day. However, that doesn't mean you have to be a professional mathematician to install a perfectly functioning plumbing system. You may not realize you are using math when you size a drain, level a water heater, or calculate a grade for a run of pipe, but you are. This module will show you the math behind common plumbing activities.

The kind of math that plumbers use is a special branch of math called **applied mathematics.** Applied mathematics is any mathematical process that you use to accomplish a specific task. It helps you get a job done and done correctly. It enables you to find the length of a run of pipe, the slope of a drain, the area of a floor, the volume of a climb-in interceptor, the temperature of wastewater—anything where you need to find a precise number. Plumbers use formulas and calculations to find those numbers. You need to understand how formulas and calculations work so that you can use them, too.

In this module, you will learn about the weights and measures that plumbers use. You will learn ways to calculate the size of flat surfaces and the amount of space inside a container. You will also learn about heat and other forces that act on liquids and gases in containers, and you will discover how plumbers use basic machines to move things. By learning about all these things, you will improve your plumbing skills, and the results will show in the quality of work you do.

2.0.0 ◆ WEIGHTS AND MEASURES

Look at the following list. What do all these things have in common?

- Weight
- Numerical and statistical data
- Time
- Angles
- Money
- Measurement
- Temperature

2.1.0 The English System

The system of weights and measures currently in use in the United States is called the **English system**. It is a version of a system created in England many centuries ago called the British Imperial system. The system was brought over by the European settlers who first established colonies in North America. The system has gradually developed over time to its current form. Some measures and values differ from the original British Imperial system. Currently, the United States is the only country that still uses this system for all its weights and measures.

Pipe, tools, fixtures, fittings, and other plumbing and construction equipment in the United States are all sized according to the English system. Many people in the United States are familiar with the weights and measures included in the English system. However, the manner in which the different measures work together (such as how many feet are in a yard) is not always known or understood. *Table 1* lists weights and measures that are commonly used by plumbers.

For one thing, all of their values can be expressed as numbers. Not only that, but they each have their own vocabulary of measurement. If you were to just say "five," no one would understand exactly what you meant. However, if you say "five *minutes*", people know you are talking about time. When you say "five *dollars*," people understand you are referring to money. Systems of measurement allow people to exchange information about all kinds of numbers. Numbers are important to plumbers. Whether it's the angle of a pipe, the amount of water in a tank, the capacity of a pump, or the distance to the sewer line, plumbers are always thinking and working in terms of how much or how little, how large or how small, and how long or how short.

When working with systems of measurement, you may be required to convert decimals to fractions. Converting decimals to fractions is a two-step process. First, multiply the decimal by 12, which is the number of inches in a foot. Then, multiply that number by the base of the fraction you are seeking. For example, multiply by 16 if you are looking for sixteenths of an inch, by 8 if you are looking for eighths of an inch, and so on.

Plumbers use two systems of measurement when working with plumbing installations. You are probably familiar with at least one of them. In this section, you will review the two major systems of weights and measures. You will also learn how to convert measurements from one system to the other.

Table 1 English System Common Weights and Measures

English System—Common Weights and Measures		
Linear Measures		
1 foot	=	12 inches
1 yard	=	36 inches
1 yard	=	3 feet
1 mile	=	5,280 feet
1 mile	=	1,760 yards
Area Measures		
1 square foot	=	144 square inches
1 square yard	=	9 square feet
1 acre	=	43,560 square feet
1 square mile	=	640 acres
Volume Measures		
1 cubic foot	=	1,728 cubic inches
1 cubic yard	=	27 cubic feet
1 freight ton	=	40 cubic feet
1 register ton	=	100 cubic feet
Liquid Volume Measures		
1 gallon	=	128 fluid ounces
1 gallon	=	4 quarts
1 barrel	=	31.5 gallons
1 petroleum barrel	=	42 gallons
Weights		
1 pound	=	16 ounces
1 hundredweight	=	100 pounds
1 ton	=	2,000 pounds

2.2.0 The Metric (SI) System

The **metric system** was developed in France in the eighteenth century. It has since become a worldwide standard in almost all professional and scientific fields. The metric system is also called the **SI system.** SI stands for *Système International d'Unités*, or International System of Units.

DID YOU KNOW?

History of the Metric System

The idea of basing weights and measures on a decimal system (numbers that are multiples of 10) goes back to the sixteenth century. In 1790, Thomas Jefferson proposed that the new United States develop a decimal weights and measures system. The same year, King Louis XVI of France instructed scientists in his country to develop such a system. Five years later, France adopted the metric system (the name comes from the Greek word *metron*, which means *measure*). In 1875, France hosted the Convention of the Metre. Eighteen nations, including the United States, signed a treaty that established an international body to govern the development of weights and measures. In 1960 this organization officially changed the name of the metric system to the International System of Units (SI).

The metric system was designed with three goals in mind:

• The system should use units based on unchanging quantities in nature.
• All measurement units should derive from only a few base units.
• The system should use multiples of 10.

The modern metric system still follows these three simple guidelines.

Originally, the meter was measured as one ten-millionth the distance from the North Pole to the equator. Bars of that length were constructed in brass and later in platinum and sent to the standards laboratories of participating countries. Scientists then learned that the actual distance from the pole to the equator was different from the measurement the French scientists used. The length of the meter was recalculated using more precise methods, including the distance traveled by a light beam in a fraction of a second.

In 1994, the United States government required that consumer products feature measurements in both SI and English form. With the advent of the International Plumbing Code® and other international standards, the United States may eventually follow the rest of the world and completely adopt the SI system of weights and measures.

One of the major advantages of the metric system is that it is rigorously standardized. This makes it easy to memorize the basic system. There are only seven basic units of measurement in the modern metric system (see *Table 2*). All other metric measures result from various combinations of these basic units. Multiples and fractions of the basic units are expressed as powers of 10 (except for seconds, of which there are 60 in a minute). A standard set of prefixes is used to denote these larger or smaller numbers (refer to *Table 2*). That way, you will always know that a kilometer is 1,000 meters just by looking at the prefix. Note that the basic measure of mass, the kilogram, is already a multiple. A kilogram is 1,000 grams. Because grams are such small amounts of mass, however, a larger base unit seemed more appropriate.

With the growing popularity of international standards, such as the International Plumbing Code®, plumbers will encounter metric weights and measures more and more in the future.

Table 2 Basics of the Metric System

The unit of...	Is called the...
Length	Meter
Mass	Kilogram
Temperature	Kelvin
Time	Second
Electric current	Ampere
Light intensity	Candela
Substance amount	Mole

The prefix...	Means...
Micro-	One millionth
Milli-	One thousandth
Centi-	One hundredth
Deci-	One tenth
Deka-	Ten times
Hecto-	One hundred times
Kilo-	One thousand times
Mega-	One million times

CAUTION

Always make sure that you express measurements using the correct system. Errors caused by using the wrong system of measurement can cost time and money. Never use a metric tool on English system pipe and fittings, or vice versa. You could damage both the tool and the fitting.

2.3.0 Converting Measurements

You've probably seen tools marked with lengths in both inches and millimeters or fixtures with shipping weights in both pounds and kilograms. However, there may be times in the field when you have to perform a calculation yourself. You should become familiar with the basic conversions between the English and metric systems.

Appendix A provides a reference for converting the most common weights and measures. To convert a measurement into its equivalent in the other system, multiply the measurement by the number in the far-right column. For example, you want to know how many kilometers equal 5 miles. Referring to the table, you see that the number must be multiplied by 1.6. The result, therefore, is that 5 miles equal 8 kilometers. Practice calculating some conversions of your own.

Review Questions

Sections 1.0.0–2.0.0

1. A mile can be divided into 5,280 feet or _____ yards.
   a. 52,800
   b. 1,760
   c. 43,560
   d. 1,728

2. One register ton equals _____ cubic feet.
   a. 27
   b. 40
   c. 100
   d. 160

3. The prefix *centi-* means _____.
   a. one hundredth
   b. one millionth
   c. one hundred times
   d. one thousand times

4. The prefix meaning *one tenth* is _____.
   a. deci
   b. deka
   c. hecto
   d. micro

5. The basic unit of light intensity in the metric system is the _____.
   a. kelvin
   b. ampere
   c. mole
   d. candela

3.0.0 ◆ MEASURING AREA AND VOLUME

Plumbers need to know how to measure the size of flat surfaces and the space inside different types of containers. What are some of the surfaces and containers you see in a typical plumbing installation? A building foundation is a flat surface, and so are floors, roofs, and parking lots. Pipes are probably the most common container you will work with as a plumber. Swimming pools, interceptors, and lavatories are also types of containers.

In this section, you will learn how to measure flat surfaces and the space inside containers. With practice, these measurements will become as familiar to you as any other tool you use on the job, like a wrench or a plumb bob. Remember that you have already learned the basic math functions covered throughout this section in *Plumbing Levels One* and *Two*.

3.1.0 Measuring Area

Area is the measurement of a flat surface (see *Figure 1*). Area is measured using two dimensions: length and width. Floor space is probably the most common type of area that plumbers encounter on the job. In *Plumbing Level Two*, you learned that plumbers must take floor space into account when designing plumbing installations. In this section you will learn how to calculate the area of floor spaces and other areas.

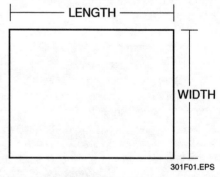

301F01.EPS

Figure 1 ◆ The concept of area.

In the English system, area is expressed in **square feet.** You will also see this written as sq. ft. or ft². In the metric system, it is measured in **square meters.** This can also be written as m².

Length and width are measured in **decimals of a foot.** You learned in *Plumbing Level One* that a decimal of a foot is a decimal fraction in which the denominator is 12, instead of the usual 10 (see *Table 3*).

Table 3 Inches Converted to Decimals of a Foot

Inches	Decimals of a Foot	Inches	Decimals of a Foot	Inches	Decimals of a Foot	Inches	Decimals of a Foot
1/16	0.005	3 1/16	0.255	6 1/16	0.505	9 1/16	0.755
1/8	0.010	3 1/8	0.260	6 1/8	0.510	9 1/8	0.760
3/16	0.016	3 3/16	0.266	6 3/16	0.516	9 3/16	0.766
1/4	0.021	3 1/4	0.271	6 1/4	0.521	9 1/4	0.771
5/16	0.026	3 5/16	0.276	6 5/16	0.526	9 5/16	0.776
3/8	0.031	3 3/8	0.281	6 3/8	0.531	9 3/8	0.781
7/16	0.036	3 7/16	0.286	6 7/16	0.536	9 7/16	0.786
1/2	0.042	3 1/2	0.292	6 1/2	0.542	9 1/2	0.792
9/16	0.047	3 9/16	0.297	6 9/16	0.547	9 9/16	0.797
5/8	0.052	3 5/8	0.302	6 5/8	0.552	9 5/8	0.802
11/16	0.057	3 11/16	0.307	6 11/16	0.557	9 11/16	0.807
3/4	0.063	3 3/4	0.313	6 3/4	0.563	9 3/4	0.813
13/16	0.068	3 13/16	0.318	6 13/16	0.568	9 13/16	0.818
7/8	0.073	3 7/8	0.323	6 7/8	0.573	9 7/8	0.823
15/16	0.078	3 15/16	0.328	6 15/16	0.578	9 15/16	0.828
1	0.083	4	0.333	7	0.583	10	0.833
1 1/16	0.089	4 1/16	0.339	7 1/16	0.589	10 1/16	0.839
1 1/8	0.094	4 1/8	0.344	7 1/8	0.594	10 1/8	0.844
1 3/16	0.099	4 3/16	0.349	7 3/16	0.599	10 3/16	0.849
1 1/4	0.104	4 1/4	0.354	7 1/4	0.604	10 1/4	0.854
1 5/16	0.109	4 5/16	0.359	7 5/16	0.609	10 5/16	0.859
1 3/8	0.115	4 3/8	0.365	7 3/8	0.615	10 3/8	0.865
1 7/16	0.120	4 7/16	0.370	7 7/16	0.620	10 7/16	0.870
1 1/2	0.125	4 1/2	0.374	7 1/2	0.625	10 1/2	0.875
1 9/16	0.130	4 9/16	0.380	7 9/16	0.630	10 9/16	0.880
1 5/8	0.135	4 5/8	0.385	7 5/8	0.635	10 5/8	0.885
1 11/16	0.141	4 11/16	0.391	7 11/16	0.641	10 11/16	0.891
1 3/4	0.146	4 3/4	0.396	7 3/4	0.646	10 3/4	0.896
1 13/16	0.151	4 13/16	0.401	7 13/16	0.651	10 13/16	0.901
1 7/8	0.156	4 7/8	0.406	7 7/8	0.656	10 7/8	0.906
1 15/16	0.161	4 15/16	0.411	7 15/16	0.661	10 15/16	0.911
2	0.167	5	0.417	8	0.667	11	0.917
2 1/16	0.172	5 1/16	0.422	8 1/16	0.672	11 1/16	0.922
2 1/8	0.177	5 1/8	0.427	8 1/8	0.677	11 1/8	0.927
2 3/16	0.182	5 3/16	0.432	8 3/16	0.682	11 3/16	0.932
2 1/4	0.188	5 1/4	0.438	8 1/4	0.688	11 1/4	0.938
2 5/16	0.193	5 5/16	0.443	8 5/16	0.693	11 5/16	0.943
2 3/8	0.198	5 3/8	0.448	8 3/8	0.698	11 3/8	0.948
2 7/16	0.203	5 7/16	0.453	8 7/16	0.703	11 7/16	0.953
2 1/2	0.208	5 1/2	0.458	8 1/2	0.708	11 1/2	0.958
2 9/16	0.214	5 9/16	0.464	8 9/16	0.714	11 9/16	0.964
2 5/8	0.219	5 5/8	0.469	8 5/8	0.719	11 5/8	0.969
2 11/16	0.224	5 11/16	0.474	8 11/16	0.724	11 11/16	0.974
2 3/4	0.229	5 3/4	0.479	8 3/4	0.729	11 3/4	0.979
2 13/16	0.234	5 13/16	0.484	8 13/16	0.734	11 13/16	0.984
2 7/8	0.240	5 7/8	0.490	8 7/8	0.740	11 7/8	0.990
2 15/16	0.245	5 15/16	0.495	8 15/16	0.745	11 15/16	0.995
3	0.250	6	0.500	9	0.750	12	1.000

You can also convert inches to their decimal equivalent in feet without the table of decimal equivalents. All you need to do is some simple division. First, express the inches as a fraction that has 12 as the denominator. You use 12 because there are 12 inches in a foot. Then reduce the fraction and convert it to a decimal.

Let's demonstrate using an example. To find the decimal equivalent in feet of 3 inches, follow these three steps:

Step 1 Express 3 inches as a fraction with 12 as the denominator.

$\frac{3}{12}$

Step 2 Reduce the fraction.

$\frac{3}{12} = \frac{1}{4}$

Step 3 Convert the reduced fraction to a decimal. You can do this by dividing the denominator, 4, into 1.00:

$$\begin{array}{r} .25 \\ 4\overline{)1.00} \\ \underline{0.8} \\ 0.20 \\ \underline{0.20} \end{array}$$

Thus, 3 inches converts to 0.25 of a foot. For more complicated fractions, use a calculator.

In plumbing, it will be useful for you to know how to calculate the area of three basic shapes:

- Rectangles
- Right triangles
- Circles

With this knowledge, you will be able to calculate most of the types of areas you will encounter on the job.

3.1.1 Measuring the Area of a Rectangle

A **rectangle** (*Figure 2A*) is a four-sided figure in which all four corners are right angles. To calculate the area (A) of a rectangle, multiply the length (l) by the width (w). Expressed as a formula, the calculation is:

A = lw

A **square** is a special type of rectangle in which all four sides are the same length. Use the same formula to determine the area of a square (see *Figure 2B*). You can also calculate the area of a square by squaring the length of one of the sides:

A = s²

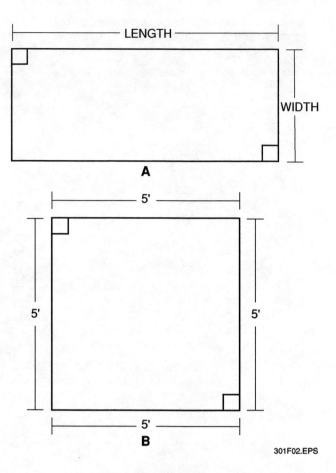

Figure 2 ◆ Measuring the area of a rectangle and a square.

Let's try this formula on a rectangle that is 5 feet 1½ inches by 8 feet 10¾ inches:

Step 1 First, convert the inches to decimals of a foot.

5 feet 1½ inches = 5.1250 feet

8 feet 10¾ inches = 8.895833 feet

Step 2 Now round your answer to three decimal places, ensuring that round-ups are accurate. When rounding numbers, any number followed by 5 through 9 is rounded up, while any number followed by 0 through 4 remains the same.

5.1250 feet = 5.125 feet

8.895833 feet = 8.896 feet

Please note the number of decimal places you round to depends on the job being done and the desired level of accuracy. When in doubt, ask you supervisor.

Step 3 Multiply the length by the width.

A = lw

A = (5.125)(8.896)

A = 45.592 sq. ft.

The total area of a rectangle that is 5 feet 1½ inches by 8 feet 10¾ inches, therefore, is 45.592 square feet.

3.1.2 Measuring the Area of a Right Triangle

A **right triangle** is a three-sided figure that has an internal angle that equals 90 degrees. The area of a right triangle is one-half the product of the base (b) and height (h). In other words:

A = ½(bh)

An **isosceles triangle** is a special kind of triangle that has two sides of equal length. Use the same formula to calculate the area of an isosceles triangle (see *Figure 3*).

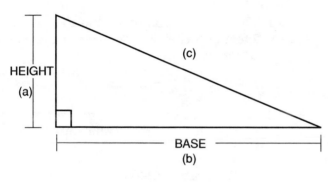

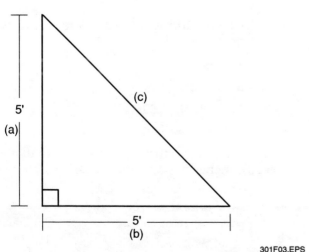

301F03.EPS

Figure 3 ◆ A right triangle and an isosceles triangle.

Test this formula on a triangle where the height is 5 feet 2¾ inches and the base is 6 feet 6½ inches. Calculate the area for this triangle by following these steps:

Step 1 Convert the inches into decimals of a foot.

5 feet 2¾ inches = 5.229166 feet

6 feet 6½ inches = 6.541666 feet

Step 2 Now round your answer to three decimal places, ensuring that round-ups are accurate.

5.229166 feet = 5.229 feet

6.541666 feet = 5.542 feet

Step 3 Determine the area of the triangle.

A = ½(bh)

A = (½)(6.542)(5.229)

A = (½)(34.208)

A = 17.104 sq. ft.

The area of a right triangle whose height is 5 feet 2¾ inches and base is 6 feet 6½ inches long, therefore, is 17.104 square feet.

Sometimes, you might not know the length of one of the legs in a right triangle. To calculate the missing leg, you can use a special calculation called the **Pythagorean theorem,** which states the following:

$a^2 + b^2 = c^2$

Refer to *Figure 3*. You see that a and b are the two sides that meet at the 90-degree angle. The side marked c is called the **hypotenuse.** To find either a or b, use the following calculations:

$a = \sqrt{c^2 - b^2}$

$b = \sqrt{c^2 - a^2}$

Use the square root function on your calculator to obtain the square root of a, b, and c. Most calculators have a square root key (see *Figure 4*). It is marked with the $\sqrt{}$ symbol. To use it, first enter the number for which you want to calculate the square root. Then press the square root key. This is the square root of the number you entered.

If the number you entered represented feet, then the decimals must be converted to inches and fractions of an inch. If the number you entered was in inches, the decimals must be converted to fractions of an inch.

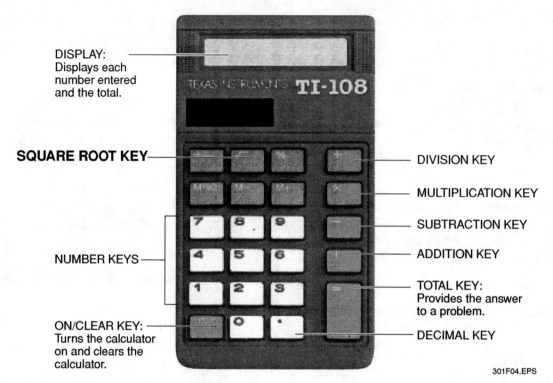

DISPLAY:
Displays each
number entered
and the total.

SQUARE ROOT KEY

NUMBER KEYS

ON/CLEAR KEY:
Turns the calculator
on and clears the
calculator.

DIVISION KEY

MULTIPLICATION KEY

SUBTRACTION KEY

ADDITION KEY

TOTAL KEY:
Provides the answer
to a problem.

DECIMAL KEY

301F04.EPS

Figure 4 ◆ Square root key.

3.1.3 Measuring the Area of a Circle

A **circle** is a figure that is bordered by a curve that keeps the same distance from the center all the way around. The strainer on a can wash drain is an example of a circular surface you will encounter as a plumber. The area of a circle is a little more complicated to calculate than that of a rectangle or triangle. To calculate the area of a circle, you will use the radius of the circle and pi. The radius of a circle is one-half the diameter (which is formed by a straight line from the center of the circle to its edge). Pi is equal to 3.1416 and is usually represented by the Greek letter π. First, multiply the radius (R) by itself (this is called *squaring*). Then, multiply the result by pi.

$$A = \pi R^2$$

To find the area of a circle with a diameter of 4 feet 3½ inches, for example, follow these steps:

Step 1 The radius is one-half the diameter, or 2 feet 1¾ inches. Now convert the inches to decimals of a foot.

2 feet 1¾ inches = 2.145833 feet

Step 2 Now round your answer to three decimal places, ensuring that round-ups are accurate.

2.145833 feet = 2.146 feet

Step 3 Use the formula to calculate the area. Do not round pi.

$A = \pi R^2$
$A = (3.1416)(2.146^2)$
$A = (3.1416)(4.605)$
$A = 14.47$ sq. ft.

You have shown that a circle with a diameter of 4 feet 3½ inches has an area of 14.47 square feet.

ON THE

· LEVEL ·

As Easy as π

The value of π represents the ratio of a circle's *circumference* to its *diameter*. The circumference is the length of a circle's curved edge. A diameter is a straight line drawn through the middle of a circle, from one side to the other. On the job, you can use 3.1416 to represent π. However, π may actually be an infinite number. Mathematicians are still not sure!

3.1.4 Area Calculations in Plumbing

You have learned how to calculate the area of different shapes. With this knowledge, you can calculate the area of almost any floor plan, no matter how simple or complex it is. Always refer to the scale in a construction drawing to determine the proper measurements of the floor.

On the job, you may have to calculate the area of a space shaped like a rectangle or a right triangle. Let's say you have to install an area drain in a parking lot that is 30 feet 6½ inches by 50 feet 3 inches. The parking lot's size affects the amount of storm water runoff that the drain will have to handle. This factor determines the size of the area drain you select.

Step 1 First, convert the inches to decimals of a foot.

30 feet 6½ inches = 30.541666 feet

50 feet 3 inches = 50.250 feet

Step 2 Now round your answer to three decimal places, ensuring that round-ups are accurate.

30.54166 feet = 30.542 feet

50.250 feet = 50.250 feet

Step 3 Multiply the length by the width.

A = lw

A = (30.542)(50.250)

A = 1,534.74 sq. ft.

The total area of a parking lot that is 30 feet 6½ inches by 50 feet 3 inches is 1,534.74 square feet.

What do you do if you have to calculate the area of a surface that is not a simple rectangle or right triangle (see *Figure 5*)? First, break the floor plan into smaller areas shaped like rectangles and right triangles. Calculate the area for each of the smaller areas using the appropriate formula. Then add all the areas together. You will have the total overall area of the floor. You were able to find it using nothing more than the equations you've just learned.

3.2.0 Measuring Volume

Volume is a measure of capacity. Put another way, volume is the amount of space within something (see *Figure 6*). Volume is measured using three dimensions: length, width, and height. That is why space is called *three-dimensional*.

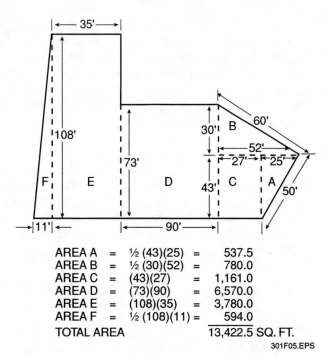

AREA A	=	½ (43)(25)	=	537.5
AREA B	=	½ (30)(52)	=	780.0
AREA C	=	(43)(27)	=	1,161.0
AREA D	=	(73)(90)	=	6,570.0
AREA E	=	(108)(35)	=	3,780.0
AREA F	=	½ (108)(11)	=	594.0
TOTAL AREA				13,422.5 SQ. FT.

301F05.EPS

Figure 5 ◆ Subdividing an area into simple shapes.

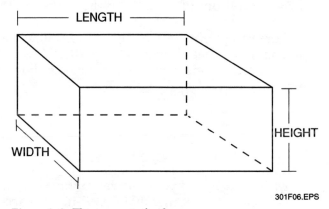

301F06.EPS

Figure 6 ◆ The concept of volume.

Volume is an important concept in plumbing. Plumbing is all about liquids moving into and out of many differently shaped spaces. Plumbers must be able to calculate the volumes of these spaces accurately to ensure the efficient operation of plumbing installations. Some of the volumes that plumbers need to be able to calculate include

- Drain receptors
- Fresh-water pipes
- Drain, waste, and vent (DWV) pipes
- Grease and oil interceptors
- Water heaters
- Septic tanks
- Swimming pools
- Catch basins

In the English system, space is measured in **cubic feet** (abbreviated as cu. ft. or ft³). The measure of liquid volume is **gallons** (abbreviated as gal.). There are 7.48 gallons in 1 cubic foot. In the metric system, the measure of space is **cubic meters** (abbreviated as m³) and the measure of liquid is in **liters** (abbreviated as L). There are 1,000 liters in a cubic meter.

You can construct three-dimensional spaces by combining flat surfaces like rectangles, squares, right triangles, and circles. When two parallel rectangles, squares, or right triangles are connected by rectangles, the space is called a **prism**. A tube with a circular cross section is called a **cylinder.**

3.2.1 Measuring the Volume of a Rectangular Prism

Rectangular prisms are three-dimensional spaces where rectangles make up the sides, top, and bottom (see *Figure 7*). A 2 × 4 piece of lumber is a rectangular prism, as is the interior of a truck bed. To calculate the length of a rectangular prism, multiply the length by the width by the height.

For example, assume you have a rectangle with a length of 5 feet 4¾ inches, a width of 2 feet 1½ inches, and a height of 6 feet. Calculate the volume this way:

Step 1 Convert the inches into decimals of a foot.

> 5 feet 4¾ inches = 5.395833 feet
>
> 2 feet 1½ inches = 2.1250 feet

Step 2 Now round your answer to three decimal places, ensuring that round-ups are accurate.

> 5.395833 feet = 5.396 feet
>
> 2.1250 feet = 2.125 feet

Step 3 Calculate the volume using the formula:

> V = lwh
>
> V = (5.396)(2.125)(6)
>
> V = (11.467)(6)
>
> V = 68.802 cu. ft.

The volume of the given rectangle is 68.802 cubic feet.

A **cube** is a rectangular prism where the length, width, and height are all the same. To find the volume of a cube, use the formula for finding the volume of a rectangular prism.

3.2.2 Measuring the Volume of a Right Triangular Prism

As you might expect, the volume of a right triangular prism is ½ times the product of the length, width, and height (refer to *Figure 7*). For example, if you had a right triangle prism with a base and width of 3 feet and a height of 5 feet, the volume would be calculated as follows:

> V = ½(bwh)
>
> V = ½(3)(3)(5)
>
> V = ½(9)(5)
>
> V = ½(45)
>
> V = 22.5 cu. ft.

The volume of the right triangular prism is 22.5 cubic feet.

3.2.3 Measuring the Volume of a Cylinder

A cylinder is a straight tube with a circular cross section (see *Figure 8*). It is a common plumbing shape. Pipes are long, narrow cylinders. Many water heaters are also shaped like cylinders. You can calculate the volume in cubic feet of a cylinder by multiplying the area of the circular base (πR²) by the height (h).

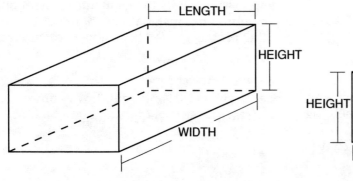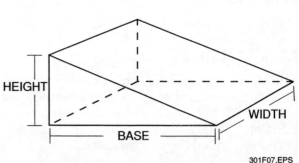

301F07.EPS

Figure 7 ◆ Rectangular prism and right triangular prism.

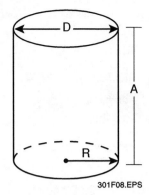

Figure 8 ◆ Cylinder.

As an example, a cylinder has an inside diameter of 30 feet and a height of 50 feet 9 inches. Obtain the volume in cubic feet using the following steps:

Step 1 The ends of a cylinder are circles. So first, you need to determine the area of the circle. In the example, the radius (half the diameter, remember) is 15 feet.

$$A = \pi R^2$$
$$A = (3.1416)(15^2)$$
$$A = (3.1416)(225)$$
$$A = 706.86 \text{ sq. ft.}$$

Step 2 To calculate the volume of the tank in cubic feet, multiply the area of the circle by the height. Remember to convert the measurement into decimals of a foot.

$$V = Ah$$
$$V = (706.86)(50.75)$$
$$V = 35,873.145 \text{ cu. ft.}$$

The volume of a tank with an inside diameter of 30 feet and a height of 50 feet 9 inches is 35,873.145 cubic feet.

3.2.4 Volume Calculations in Plumbing

For manufactured items, you can usually find the dimensions and volume in the manufacturer's specifications. For site-built containers, such as large grease interceptors, you will have to calculate the volume yourself. Don't forget to calculate the volume of objects inside a container, such as baffles, fittings, or even fixtures. Calculate their volumes and *subtract* them from the overall volume of the space (see *Figure 9*).

Remember that when determining volume it is important also to consider weight. Take, for example, a swimming pool installed on a patio. The larger the pool, the more water it will hold and the heavier the water inside it will be. Consequently,

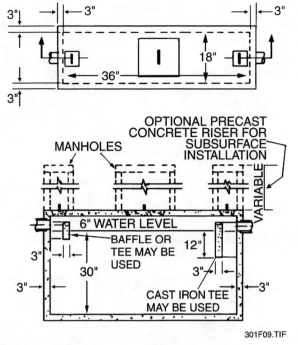

Figure 9 ◆ Calculating the volume of a site-built grease interceptor.

the deck will require more support to hold the pool. When installing pipes and fixtures, plumbers need to be able to calculate the weight of water for a given volume to ensure that the pipe or fixture is adequately supported. Local codes provide requirements on proper support, but it is important for plumbers to understand why codes do that. For example, given that a cubic foot of water weighs 62.3 pounds, how much will the water in a 200-foot section of a 4-inch water line weigh?

Once you have calculated the volume of a container such as a pipe, water tank, slop sink, or interceptor, you need to perform one more step to determine its gallon capacity. One gallon of water weighs 8.33 pounds. There are 7.48 gallons in a cubic foot. To find the capacity in gallons of a volume measured in cubic feet, multiply the cubic footage by 7.48. If you multiply the weight of 1 gallon of water by the number of gallons in a cubic foot, for example, you will find that a cubic foot of water weighs 62.3 pounds. Remember that these weights are for pure water. Wastewater will have other liquids and solids in it. This means the weight of a volume of wastewater may be greater than that of fresh water. To convert measurements in cubic inches to gallons, divide the cubic inch measurement by 231.

When installing water supply lines, using a smaller pipe size can save the customer both energy and money. You can calculate the difference using the above formula. For example, a ¾-inch-diameter pipe has a cross-sectional area of 0.442 square inches, and a ½-inch-diameter pipe has a cross-sectional area of 0.196 square inches (see *Figure 10*). If you perform the calculations, you will find that a ¾-inch pipe will hold 2¼ times the amount of water that a ½-inch pipe will hold. With a larger pipe, a faucet will have to run longer to get hot water to a sink. In addition, the amount of hot water that will be left in the pipe to cool once the faucet is turned off is much greater with a larger pipe. Always consult the local codes to determine the water- and energy-saving measures possible for each job.

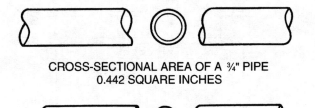

CROSS-SECTIONAL AREA OF A ¾" PIPE
0.442 SQUARE INCHES

CROSS-SECTIONAL AREA OF A ½" PIPE
0.196 SQUARE INCHES

301F10.TIF

Figure 10 ◆ Cross-sectional areas of pipe.

You can calculate the volume of a complex shape by calculating the volume of its various parts. For example, divide the shape in *Figure 11* into a half cylinder and a rectangular solid.

The volume of the rectangular portion is found as follows:

$V = lwh$

$V = (12)(6)(8)$ feet

$V = 576$ cu. ft.

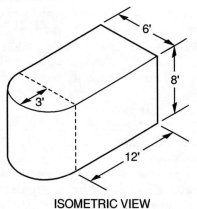

ISOMETRIC VIEW

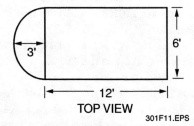

TOP VIEW

301F11.EPS

Figure 11 ◆ Complex volume.

The volume of the half cylinder is found as follows:

$V = (\pi R^2 h) \div 2$

$V = ((3.1416)(3)(3)(8)) \div 2$

$V = (226.1952) \div 2$

$V = 113.098$ cu. ft.

Now add the two together to get the total volume:

$V = 576$ cu. ft. $+ 113.098$ cu. ft.

$V = 689.098$ cu. ft.

Review Questions

Section 3.0.0

Remember, when necessary, to round your answers to three decimal places.

1. The number 7 expressed in decimals of a foot is _____.
   a. 84
   b. 0.583
   c. 0.84
   d. 0.292

2. The area of a rectangle with a length of 3 feet and a width of 6 feet is _____.
   a. 9 square feet
   b. 36 square feet
   c. 18 square feet
   d. 8.48 square feet

3. The area of a circle with a diameter of 4 inches is _____.
   a. 16 square inches
   b. 157.974 square inches
   c. 33.333 square inches
   d. 12.566 square inches

4. The volume of a right triangular prism with a base of 2 feet, a width of 5 feet, and a height of 7 feet is _____.
   a. 70 cubic feet
   b. 14 cubic feet
   c. 35 cubic feet
   d. 122 cubic feet

5. The volume of a cylinder with a diameter of 30 inches and a height of 60 inches is _____.
   a. 42,411.60 cubic inches
   b. 169,646.40 cubic inches
   c. 13,324.037 cubic inches
   d. 706.86 cubic inches

4.0.0 ◆ TEMPERATURE, PRESSURE, AND FORCE

In the previous section, you learned how to calculate the space inside three-dimensional containers like pipes, interceptors, and swimming pools. Those calculations are an important step when installing a plumbing system. But once a plumbing system is in place, it has to contain and move liquids and gases. Liquids, such as water, and gases, such as air, have properties that affect how they behave inside the system. In this section, you will learn about three of the most important of these properties: **temperature**, **pressure**, and force.

Temperature, pressure, and force are very important concepts in plumbing. Pipes, fittings, and fixtures are all designed to work within a range of acceptable temperatures and pressures. If those limits are exceeded, the system could be severely damaged, and people could be injured as a result. A plumbing system that contains materials not appropriate for its operating conditions is also in danger of failing and causing damage and injury. Every plumbing system is designed from the outset with these considerations in mind, so every plumber needs to know the principles that govern temperature and pressure.

DID YOU KNOW?

Microwave Ovens— The Power to Move Molecules

Microwave ovens are a common kitchen appliance. People use microwaves for everything from thawing frozen food to boiling a cup of water and cooking a steak. Microwave ovens cook food faster and more efficiently than conventional ovens. How do they work? Microwaves are tiny radio waves that are absorbed by water molecules, fats, and sugars. Other materials, such as plastic or ceramic, do not absorb microwaves. When microwaves are absorbed, they convert directly into heat.

When a molecule is heated, scientists say it becomes *excited.* Heated molecules move fast, bounce against one another, and ricochet off container walls. Though scientists can't actually see individual molecules, they can see the effect of their excited state. When heated molecules bump into a larger particle, the molecules cause the particle to move in a zigzag pattern. This form of motion is called Brownian motion, in honor of Robert Brown, a scientist who mathematically described it in 1827.

4.1.0 Temperature

Temperature is a measure of the warmth or coldness of an object according to a scale. Heat transfers from warmer objects to cooler ones. This is because heat is a form of energy, and energy seeks **equilibrium.** Equilibrium is a condition in which the temperature is the same throughout the object or space. This is why you get hot when you are working next to a steam pipe; your body is absorbing some of the heat energy of the pipe. The flow of energy from a hotter object to a cooler one is called **conduction.**

Plumbers are very concerned with heat flow when they install a plumbing system. Pipes can lose heat into the surrounding atmosphere, into a nearby wall or ceiling, or into the ground. When the pipe loses too much heat, it freezes. On the other end of the scale, plumbers are also concerned about how much heat energy enters a plumbing system. Too much heat can also cause damage. That is why, for example, hot wastes are allowed to cool before they enter the sanitary system.

4.1.1 Thermometers

Thermometers are tools that measure temperature. When you think of a thermometer, chances are the image you have is of a long glass tube that you use to check your temperature when you are sick. However, there are as many different shapes of thermometers as there are applications. Thermometers are installed wherever there is a need to know a temperature range. Always consult the manufacturer's instructions before installing or using a thermometer.

There are three popular types of thermometers. **Liquid thermometers** use a glass tube filled with fluid, such as mercury or alcohol. The fluid expands when heated and rises in the tube. Lines on the tube corresponding to the height of the fluid indicate the temperature. Liquid thermometers are commonly used in refrigerators and freezers (see *Figure 12*). **Bimetallic thermometers** use **thermal expansion** to show a temperature reading. Inside a bimetallic thermometer is a coil of metal. The coil is made up of two thinner strips of metal bonded together. Each of these strips expands at a different rate when heated. This causes the coil to curve when the thermometer is placed near a heat source.

301F12.EPS

Figure 12 ◆ Liquid thermometer for refrigerators and freezers.

An indicator attached to the coil then points to temperature lines marked on a dial (see *Figure 13*). Bimetallic thermometers are found in thermostats, for example, in home and commercial heating/cooling systems. **Electrical thermometers** convert electrical resistance or voltages generated by heat into a temperature reading. Electrical thermometers usually have a digital readout (see *Figure 14*). Electrical thermometers are used in weather balloons and in digital bank clocks.

301F13.EPS

Figure 13 ◆ Bimetallic thermometers.

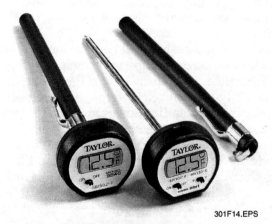

301F14.EPS

Figure 14 ◆ Electrical thermometers.

 WARNING!

Water at 110°F and above can scald on contact. Wear appropriate personal protective equipment and observe proper safety procedures when working around hot water sources.

Thermometers usually display temperatures using the **Fahrenheit scale,** the **Celsius scale,** or sometimes both. These two scales are widely used in commercial and industrial thermometers. On the Fahrenheit (F) scale, 32°F is the freezing point of pure water at sea level. The boiling point of pure water is 212°F.

Unlike the Fahrenheit scale, the Celsius scale is deliberately calibrated to the temperature of pure water at sea level pressure. On the Celsius (C) scale, therefore, 0°C is the freezing point of pure water and 100°C is the boiling point. The Celsius scale is a **centigrade scale** because it is divided into 100 degrees (you learned earlier that *centi-* means *one hundredth*).

The **Kelvin scale** is the basic system of temperature measure in the metric system. Its zero point is the coldest temperature that matter can attain in the universe, according to science (see *Figure 15*). Beginning at 253 K, the all-positive scale goes up in 10-degree increments. The Kelvin scale is the measurement system used by plumbers. For example, rated temperatures in manufacturer's specifications for fittings and fixtures are provided in degrees Kelvin.

Centigrade (Celsius)	Fahrenheit	Kelvin	
°C	°F	K	
100	212	373	WATER BOILS
90	194	363	
80	176	353	
70	158	343	
60	140	333	
50	122	323	
40	104	313	
30	86	303	
20	68	293	
10	50	283	
0	32	273	WATER FREEZES
−10	14	263	
−20	−4	253	

301F15.EPS

Figure 15 ◆ Fahrenheit, Celsius, and Kelvin temperature scales.

To convert Fahrenheit measurements into Celsius measurements, first subtract 32 from the Fahrenheit temperature, then multiply the result by the fraction ⅝. The formula can be written this way:

$$C = (F{-}32) \times \tfrac{5}{9}$$

For example, let's suppose you wanted to find the Celsius equivalent to 140°F. Apply the following formula:

$$C = (140{-}32) \times \tfrac{5}{9}$$
$$C = (108) \times \tfrac{5}{9}$$
$$C = \tfrac{540}{9}$$
$$C = 60$$
140°F equals 60°C.

To find the Fahrenheit equivalent of a Celsius temperature, what do you think the formula will be? First, multiply the Celsius number by the fraction ⅝. Then add 32. The formula looks like this:

$$F = (C \times \tfrac{9}{5}) + 32$$

DID YOU KNOW?

Fahrenheit, Celsius, and Kelvin

The Fahrenheit, Celsius, and Kelvin scales are named after pioneering scientists who helped develop modern methods of temperature measure.

Daniel Gabriel Fahrenheit (1686–1736) was a German physicist. In 1709, he invented a liquid thermometer that used alcohol. Five years later, he invented the mercury thermometer. In 1724, he developed a temperature scale that started at the freezing point of salt water. This is the modern Fahrenheit scale.

Anders Celsius (1701–1744) was a Swedish astronomer and professor who developed a centigrade scale in 1742. He based the scale's zero-degree mark on the freezing point of pure water and the 100-degree mark on the boiling point of pure water. The modern centigrade temperature scale is the same one that Celsius developed, and it is named in his honor.

William Thomson, Lord Kelvin (1824–1907), was a British physicist and professor. He conducted pioneering research in electricity and magnetism. He also participated in a project to lay the first communications cable across the Atlantic in 1857. Kelvin was also an inventor. In the mid-nineteenth century, he developed a measuring system based on the Celsius scale. The zero point of the scale is the theoretical lowest temperature in the universe. The Kelvin scale is the metric system's standard temperature scale.

To test the formula, practice converting 10°C into Fahrenheit:

$$F = (10 \times \tfrac{9}{5}) + 32$$
$$F = (\tfrac{90}{5}) + 32$$
$$F = 18 + 32$$
$$F = 50$$
10°C equals 50°F.

4.1.2 Thermal Expansion

Materials expand when they are heated and contract when they cool. The change in size caused by heat is called *thermal expansion*. Thermal expansion occurs in all the dimensions of an object: length, width, height, and even thickness. Thermal expansion is affected by several factors:

- *Length of the material*—the longer the material, the greater the increase in the length.
- *Temperature change*—the greater the temperature, the greater the increase in size.
- *Type of material*—different materials expand at different rates (see *Figure 16*).

Thermal expansion alters the area and volume of the heated material. For pipe, that means diameter, thread size and shape, joints, bends, and length are all affected. These changes could pose a serious problem for the safe and efficient operation of a plumbing system. Always use materials that are appropriate for the temperature range. Consult your local code or talk to a local code expert about temperature considerations in your area.

4.1.3 Protecting Pipes From Freezing

Plumbers install pipes, fittings, fixtures, and appliances so that they don't suffer from (or cause) repeated expansion and contraction. Underground pipes that are at risk of freezing are installed below the frost line. Depending on where in the country you work, the frost line depth may vary. Consult the local code and building officials. Remember that codes require accessibility to valves on water service pipes and cleanouts on the sanitary sewer drainpipe.

Snow acts like insulation and keeps frost from penetrating the soil too deeply. Therefore, the frost depth on snow-covered ground is significantly less than it is in other areas that are routinely cleared of snow, such as driveways. Therefore, in colder climates, do not locate the service entrance under a sidewalk, driveway, or patio. Note also that frost penetrates wet soils more deeply than it does dry soils. If possible, locate water pipes under lawns or other areas where snow will remain. Inside buildings, use insulation to protect pipes from freezing.

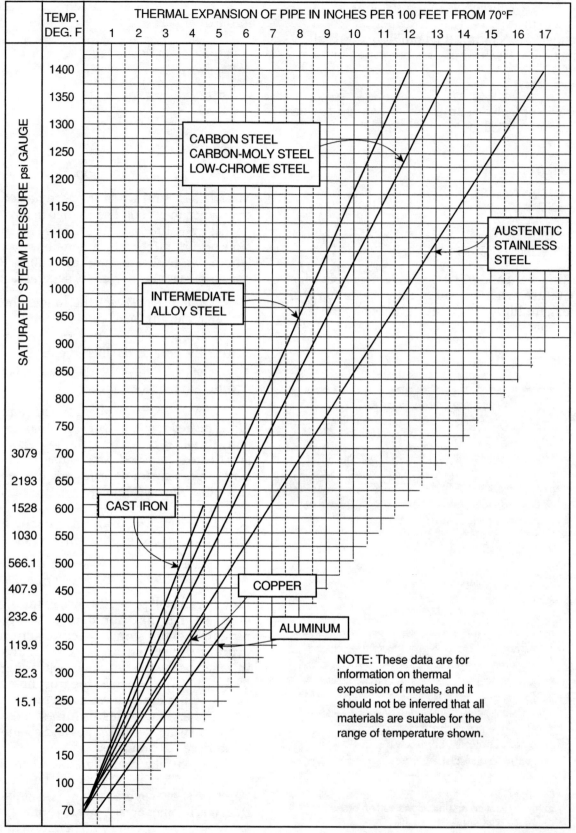

Figure 16 ◆ Thermal expansion of different types of pipe metals.

4.2.0 Pressure and Force

How can a slug of air in a drainpipe blow out a fixture trap seal? Why does water in a tank high aboveground flow with more force than water in a tank at ground level? The answer to both questions has to do with pressure. Pressure is the force created by a liquid or gas inside a container. When a liquid or gas is compressed, it exerts more pressure on the surrounding environment. If the compressed liquid or gas finds an outlet, it will escape. The speed of the escaping liquid or gas is proportional to the amount of pressure to which it was subjected. In the examples above, the pressure applied by the weight of a water column causes both reactions.

In the English system of measurement, pressure is measured in **pounds per square inch** (psi). There are two ways to measure psi:

- *Pounds per square inch absolute (psia)*—uses a perfect vacuum as the zero point. The total pressure that exists in the system is called the *absolute pressure.*

- *Pounds per square inch gauge (psig)*—uses local air pressure as the zero point. Gauge pressure is pressure that can be measured by gauges.

You can see that psig is a relative measure. Absolute pressure is equal to gauge pressure plus the atmospheric pressure. That's why if you are in Miami (at sea level), 0 psig will be 14.7 psia. However, if you are in Denver (elevation 5,280 feet), 0 psig will be 12.1 psia. The relationship between the two forms of measurement is easy to remember:

psia = psig + local atmospheric pressure

Pressure is related to temperature; increasing the pressure in a container will increase the temperature of the liquid or gas inside. Likewise, by increasing the temperature of a liquid or gas in a container, the pressure will rise. This is why steam whistles out of a kettle when you boil water on the stove. If the steam did not escape, the pressure would eventually blow off the lid of the kettle instead. In the case of excess temperature and pressure in a 40-gallon water heater, the consequences would be much more severe.

Pressure is a fundamental concept in plumbing. Two of the most important forms of pressure in plumbing are air pressure and water pressure. Atmospheric air is a gas, and therefore it exerts pressure. Air at sea level (an altitude of zero feet) creates 14.7 pounds of pressure per square inch of surface area. Residential and commercial plumbing installations operate at this pressure. Vents maintain air pressure within a plumbing system. Vents allow outside air to enter and replace wastewater flowing out of various parts of the system. You learned how to install vents in *Plumbing Level Two.*

4.2.1 Water Hammer

When liquid flowing through a pipe is suddenly stopped, vibration and pounding noises, called *water hammer*, result. Water hammer is a destructive form of pressure. The forces generated at the point where the liquid stops act like an explosion. When water hammer occurs, a high-intensity pressure wave travels back through the pipe until it reaches a point of relief. The shock wave pounds back and forth between the point of impact and the point of relief until the energy dissipates (see *Figure 17*). Installing water hammer arresters prevents this type of pressure from reducing the service life of pipes.

ON THE

· LEVEL ·

Pressure Versus Force

If you have two vertical pipes of different diameters, cap the bottom ends of both, and fill them each with water up to the same height (for example, 2 feet of water). The pressure on the cap will be the same in both pipes, but the force of the water will be greater in the wider pipe because there is more water in the pipe.

Pressure is a function of height, not volume, and force is a measure of volume, not height. The reason a higher column of water makes water flow faster through a spigot has nothing to do with the amount of water in the pipe, but has to do with the height of the column.

Think of it this way: take 15 cubic feet of water and put it in a 4-inch pipe. Call the amount of pressure "x" and the amount of force "y." If you took that 15 cubic feet of water and put it in a pipe that was only 2 inches in diameter, the pressure would still be "x" (because the amount of water is the same), but the force would now be "2y" (because the column of water is twice as high).

Plumbers need to know pressure to make sure the fixtures work the way they are designed to and force to make sure the pipes they installed on those fixtures don't break.

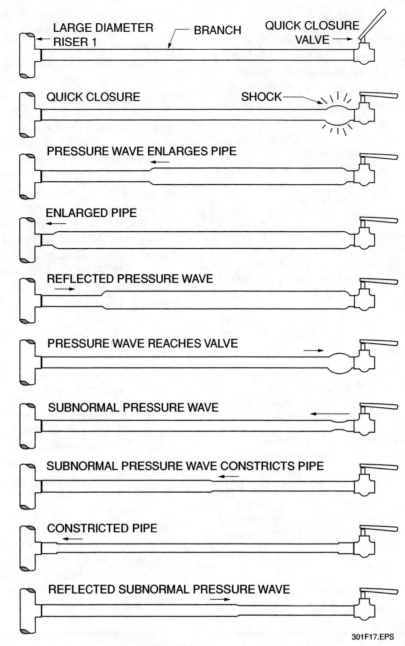

Figure 17 ◆ Water hammer.

4.2.2 Head

Water pressure is measured by the force exerted by a water column. A water column can be a well, a run of pipe, or a water tower. The height of the water column is called **head** (*Figure 18*). Head is measured in feet. The pressure exerted by a rectangular column of water 1 inch long, 1 inch wide, and 12 inches (1 foot) high is .433 psig. Another way to express the same measurement is to say that 1 foot of head is equal to .433 psig. Each foot of head, regardless of elevation, exerts .433 psig. It takes 27.648 inches of head to exert 1 psig.

The pressure, however, is determined solely by vertical height. Fittings, angles, and horizontal runs do not affect the pressure exerted by a column of water. You can use a simple formula to calculate psi:

$$psi = elevation \times .433$$

Use this formula to determine the pressure required to raise water to a specified height and the pressure created by a water column from an elevated supply. For example, if you wanted to determine the pressure required to raise water to the top of a 50-foot building, perform the following calculation:

$$psi = 50 \times .433$$
$$psi = 21.65$$

To raise a column of water 50 feet, therefore, you need to supply 21.65 psi. If you want to determine the pressure created by a 30-foot head of water in a plumbing stack, you would calculate the following:

$$psi = 30 \times .433$$

You would find that the column of water is creating a force of 12.99 psi on the bottom of the stack.

Consider, for example, a six-floor building with a 150-foot-high tank (see *Figure 18*). Note the decrease in pressure as each floor gets higher. You need a force of 43.3 psi on the bottom of the stack, but to raise the column of water to the sixth floor, you need to supply 21.7 psi.

4.2.3 Calculating Force on Test Plugs

Plumbers are responsible for arranging proper testing of drain, waste, and vent (DWV) systems and water supply piping. Water and air tests are the most common ways to ensure compliance with plans and codes. Inspectors block pipe openings with test plugs and increase the force in the system to test for leaks (see *Figure 19*). You can calculate the force applied to test plugs using the mathematical formulas you have learned in this module.

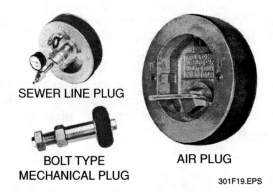

SEWER LINE PLUG

BOLT TYPE
MECHANICAL PLUG

AIR PLUG

301F19.EPS

Figure 19 ◆ Examples of air and mechanical test plugs.

Calculating these forces will help ensure that the plumbing system can withstand operating forces. For example, if you want to determine the total force on a 4-inch-diameter test plug with a water head of 25 feet, follow these steps:

Step 1 Calculate the force of the water.

$$psi = 25 \times .433$$
$$psi = 10.825$$

Step 2 Calculate the area of the test plug.

$$A = \pi \times 2^2$$
$$A = 3.1416 \times 4$$
$$A = 12.566 \text{ sq. in.}$$

Step 3 Solve for the force on the plug.

$$psig = (10.825)(12.566)$$
$$= 136.031$$

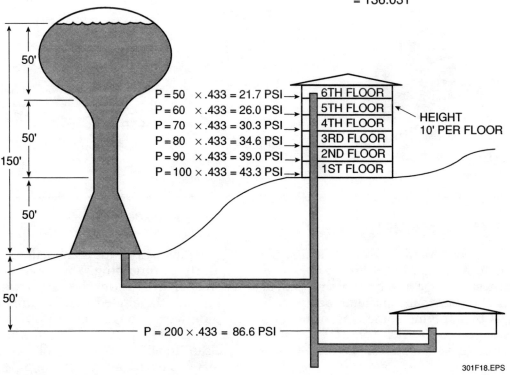

P = 50 × .433 = 21.7 PSI → 6TH FLOOR
P = 60 × .433 = 26.0 PSI → 5TH FLOOR
P = 70 × .433 = 30.3 PSI → 4TH FLOOR
P = 80 × .433 = 34.6 PSI → 3RD FLOOR
P = 90 × .433 = 39.0 PSI → 2ND FLOOR
P = 100 × .433 = 43.3 PSI → 1ST FLOOR

HEIGHT 10' PER FLOOR

50'
50'
150'
50'
50'

P = 200 × .433 = 86.6 PSI

301F18.EPS

Figure 18 ◆ A water storage tank supplying water to a six-story building.

The force on a 4-inch-diameter test plug from a water head of 25 feet is 136.03 psig. The proper head is the pressure required by the local code for a pressure test. The larger the diameter of a vertical pipe, the larger the force exerted on the pipe walls, fixtures, and plugs. Although the pressure is the same as for a narrower pipe of the same height, force varies with the volume.

4.2.4 Temperature and Pressure in Water Heaters

Water under pressure boils at a higher temperature than water at normal pressure. At normal (sea level) pressure, water boils at 212°F. However, at 150 pounds per square inch (psi), water boils at 358°F. You are already familiar with special valves that control temperature and pressure in water heaters (see *Figure 20*). Temperature regulator valves operate when the water gets too hot. Pressure regulator valves operate before the pressure of the water becomes too strong for the water heater to withstand. Combination temperature and pressure (T/P) relief valves prevent damage caused by excess temperatures and pressures. Codes require the installation of T/P relief valves on water heaters. Refer to your local code for specific guidelines. Always follow the manufacturer's instructions to ensure that the heater functions within specified operating temperatures and pressures.

301F20.EPS

Figure 20 ◆ Relief valve.

5.0.0 ◆ SIMPLE MACHINES

Machines and people perform **work**. Whether the work is moving a car on wheels down a road, digging a trench, spinning a drill bit at high speed, or lifting a pallet of materials, machines and people move things by applying some kind of force to them. Muscle power, electricity, air or water pressure, and engines can all supply the force to run machines.

DID YOU KNOW?
The idea of work is common sense, but here is the concept that's behind it. Work is what happens when force is used to move a load over a distance. The amount of work (W) can be calculated by multiplying force (F) by distance (D): $W = F \times D$.

Machines perform work by combining different types of actions or by performing just one action over and over again. A machine that performs a single action is called a **simple machine.**

There are six types of simple machines:

- The **inclined plane**
- The **lever**
- The **pulley**
- The **wedge**
- The **screw**
- The **wheel and axle**

A complex machine is formed when two or more simple machines are combined.

5.1.0 Inclined Planes

An inclined plane is a straight, slanted surface (see *Figure 21*). Loads can be raised or lowered along an inclined plane. Ramps are a common type of inclined plane. The longer the inclined plane is, the easier it is to move a load up or down. In other words, with an inclined plane, greater distances require less force.

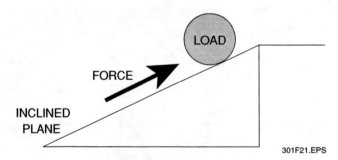

Figure 21 ◆ Inclined planes.

A pipe with grade is an inclined plane used regularly in plumbing. You have learned that the steeper the grade, the faster the liquid flows through the angled pipe. Plumbers must install pipe with the correct grade. Too much grade means that liquids flow too fast to scour the pipe clean. Too little grade will allow deposits to settle that will eventually block the pipe.

5.2.0 Levers

A lever is a straight bar, such as a pole or a board, that is free to pivot around a hinge or a support. The hinge or support is called a **fulcrum.** Operate a lever by applying force to the bar. This causes the bar to rise, carrying a load with it. The longer the lever is (that is, the farther the force is from the fulcrum), the easier it is to move the load. There are three types, or classes, of levers (see *Figure 22*):

- *First-class lever*—The fulcrum is located in the middle of the bar. The force is applied in a downward motion to one side of the fulcrum, and the load is on the opposite side of the fulcrum. A playground seesaw is a familiar type of first-class lever.
- *Second-class lever*—The load is between the fulcrum and the force. The force is applied upward. This type of lever requires much less force to move a load. However, the load does not move as far as when a first-class lever is used. Float arms, used on float-controlled valves in toilets, and wheelbarrows are examples of a second-class lever.

- *Third-class lever*—The force is applied between the fulcrum and the load. This type of lever requires the most force to move the load. However, the distance traveled by the load is also the greatest of the three types of levers. A leaning ladder that is pushed up into place from underneath is an example of a third-class lever. The ladder is the load, the feet are the fulcrum, and the pushing motion is the force.

A door is a vertical second-class lever in which the fulcrum is the hinge. Many tools are levers. Wrenches and pliers are good examples. Wrenches are second-class levers because the load being moved is close to the fulcrum. In fact, if you look closely, you will see that the load *is* the fulcrum (see *Figure 23*). Pliers are actually two levers put together. When you use a claw hammer to pry a nail out of wood, you are using a lever. Identify what class of lever a claw hammer is.

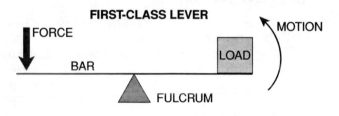

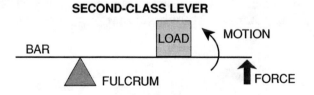

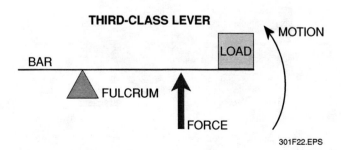

Figure 22 ◆ The three classes of levers.

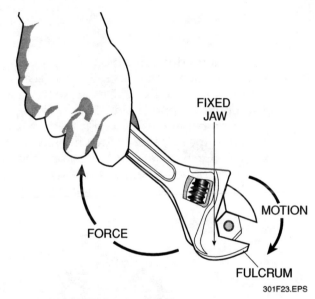

Figure 23 ◆ A wrench is a type of lever.

5.3.0 Pulleys

A pulley is a rope wrapped around a wheel (see *Figure 24*). When the rope is pulled or released, the object being lifted also moves. Pulleys and sets of pulleys both work in two different ways:

- They change the *direction* of the force—for example, pulling *down* on the rope raises the load *up*.
- They change the *amount* of force—by combining several pulleys, a lesser force can move a heavier load.

A chain could be used to turn a gate valve that is high off the ground, creating a complex pulley.

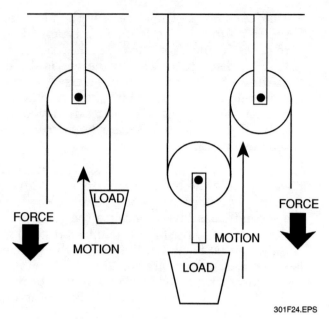

Figure 24 ◆ Pulleys.

5.4.0 Wedges

A wedge is a modified inclined plane. It is made up of two inclined planes back-to-back (see *Figure 25*). Wedges are used to cut or separate objects, prop something up, or hold something in place. You have probably used a wedge to help level a board, hold pipe or batter boards in place, or split wood. A wedge transfers the force along the same line as it is applied. Plumbers use wood chisels and cold chisels, both of which are types of wedges.

Figure 25 ◆ Types of wedges.

5.5.0 Screws

Like the wedge, the screw is a modification of an inclined plane. A screw is really a very long inclined plane that has been wrapped around a cylindrical shaft. When you apply twisting force to a screw perpendicular to the shaft, the plane transforms that into a motion that is parallel to the shaft. Because the inclined plane on a screw is so long, you need less force to move the screw.

Plumbers use screws to fasten pipe clamps and hangers to studs, assemble fixtures and appliance components, and drill wells. Adjustable wrenches use screws to widen and narrow the jaws to fit around a nut. Drill bits also use the screw principle (see *Figure 26*).

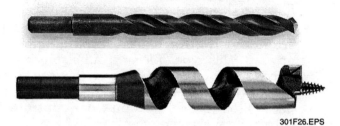

Figure 26 ◆ Drill bits are a type of screw.

5.6.0 Wheels and Axles

The wheel and axle is a modification of the basic pulley concept. The wheel is a circle that is attached to a shaft or post, called an *axle*, that runs through the center. The wheel and axle turn together (see *Figure 27*). The wheel is probably the most common simple machine. Many tools have wheels and axles. Wheelbarrows use wheels to roll loads along the ground. The handwheel and stem on certain types of valves are another familiar use of the wheel and axle combination. List some other types of wheels that plumbers use.

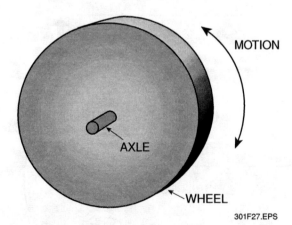

Figure 27 ◆ Wheel and axle.

Review Questions

Sections 4.0.0–5.0.0

1. The _____ scale is the standard metric temperature scale.
   a. Fahrenheit
   b. ampere
   c. mole
   d. Kelvin

2. To convert a Fahrenheit reading to Celsius, you would use the formula _____.
   a. $C = (F \times \frac{9}{5}) + 32$
   b. $C = (F + 32) \times \frac{9}{5}$
   c. $C = (F - 32) \times \frac{5}{9}$
   d. $C = (F - 32) \times \frac{9}{5}$

3. To find the pressure of a column of water, multiply the height of the column by _____.
   a. .433
   b. 3.1416
   c. 14.7
   d. .833

4. The pressure of a water head of 5 feet on a test plug with a diameter of 6 inches is _____.
   a. 41.283 psi
   b. 141.372 psi
   c. 61.214 psi
   d. 244.856 psi

5. In a third-class lever, _____.
   a. the load is located between the fulcrum and the force
   b. the force is applied between the fulcrum and the load
   c. the force is applied outside the fulcrum and load
   d. the fulcrum is located between the force and the load

6.0.0 ◆ THE WORKSHEET

Refer to the appropriate sections in the module to answer the following questions. Remember to show all your work.

1. Convert 3.5 cubic yards to cubic feet.

   _____

2. How many centimeters are in 6.5 dekameters?

   _____

3. How many fluid ounces are in 75 grams?

   _____

4. What is the area of a rectangular roof that is 25 feet 4½ inches by 60 feet 3½ inches?

   _____

5. Refer to Appendix C. Determine the area of this shape.

   _____

6. What is the volume of a rectangular prism with a length of 6 feet 3 inches, a width of 2 feet 2¼ inches, and a height of 4 feet?

   _____

7. What is the equivalent temperature in Fahrenheit of 129 degrees Celsius?

   _____

8. What is the pressure of a water head of 75 feet on a test plug with a diameter of 6 inches?

   _____

9. Draw a second-class lever. Show the bar, force, fulcrum, and load.

   _____

10. In what two ways can a pulley work?

Summary

Applied mathematics—math that you use to accomplish a specific task—is an essential part of plumbing. Plumbers use math in almost every step of plumbing installation, from design to construction to testing. In this module, you learned about the English and metric systems of measurement. Plumbers use these systems to express how high, how heavy, how much, and how many. You learned how to calculate the area of flat surfaces, such as squares, rectangles, and circles. You also reviewed the formulas for calculating the volume of prisms and columns. You were introduced to the concepts of temperature and pressure and how they affect plumbing installations, and you discovered the mechanical principles behind the tools that you use on the job.

Applied mathematics helps plumbers get a job done, and done correctly. It is as important as any other tool that plumbers use. Take the time to master the basic weights, measures, and formulas. Knowledge of applied mathematics is an essential part of your professional development.

DID YOU KNOW?
The Simplicity of Complex Machines

A complex machine is made up of two or more simple machines. Two like simple machines can be combined—for example, one lever can be used to move another lever. Combining different types of simple machines leads to common tools we see everywhere. Wheelbarrows use levers and wheels and axles. Backhoes and cranes both use levers, pulleys, and wheels and axles. Combing tools also allows work to be completed more efficiently. For example, by adding more pulleys, heavier loads can be lifted with less effort. This is called *mechanical advantage*.

Trade Terms Introduced in This Module

Applied mathematics: Any mathematical process used to accomplish a task.

Area: A measure of a surface, expressed in square units.

Bimetallic thermometer: A *thermometer* that determines *temperature* by using the *thermal expansion* of a coil of metal consisting of two thinner strips of metal.

Celsius scale: A *centigrade scale* used to measure *temperature*.

Centigrade scale: A scale divided into 100 degrees. Generally used to refer to the metric scale of temperature measure (see *Celsius scale*).

Circle: A surface consisting of a curve drawn all the way around a point that keeps the same distance from that point.

Conduction: The transfer of heat energy from a hot object to a cool object.

Cube: A rectangular prism in which the lengths of all sides are equal. When used with another form of measurement, such as *cubic meter*, the term refers to a measure of volume.

Cubic foot: The basic measure of *volume* in the *English system*. There are 7.48 *gallons* in a cubic foot.

Cubic meter: The basic measure of *volume* in the *metric system*. There are 1,000 *liters* in a cubic meter.

Cylinder: A pipe- or tube-shaped space with a circular cross section.

Decimal of a foot: A decimal fraction where the denominator is either 12 or a power of 12.

Electrical thermometer: A *thermometer* that measures *temperature* by converting heat into electrical resistance or voltage.

English system: One of the standard systems of weights and measures. The other system is the *metric system*.

Equilibrium: A condition in which all objects in a space have an equal temperature.

Fahrenheit scale: The scale of temperature measurement in the *English system*.

Fulcrum: In a *lever*, the pivot or hinge on which a bar, a pole, or other flat surface is free to move.

Gallon: In the *English system*, the basic measure of liquid volume. 7.48 gallons equal one *cubic foot*.

Head: The height of a water column, measured in feet. One foot of head is equal to .433 *pounds per square inch* gauge.

Hypotenuse: In a *right triangle*, the side opposite the right angle.

Inclined plane: A straight and slanted surface.

Isosceles triangle: A triangle in which two of the sides are of equal length.

Kelvin scale: The scale of temperature measurement in the metric system.

Lever: A simple machine consisting of a bar, a pole, or other flat surface that is free to pivot around a *fulcrum*.

Liquid thermometer: A *thermometer* that measures *temperature* through the expansion of a fluid, such as mercury or alcohol.

Liter: In the *metric system*, the basic measure of liquid volume. 1,000 liters equal one *cubic meter*.

Metric system: A system of measurement in which multiples and fractions of the basic units of measure are expressed as powers of 10. Also called the *SI system*.

Pounds per square inch: In the *English system,* the basic measure of *pressure.* Pounds per square inch (psi) is measured in pounds per square inch absolute (psia) and pounds per square inch gauge (psig).

Pressure: The force applied to the walls of a container by the liquid or gas inside.

Prism: A *volume* in which two parallel rectangles, squares, or right triangles are connected by rectangles.

Pulley: A *simple machine* consisting of a rope wrapped around a wheel.

Pythagorean theorem: A mathematical formula for finding the *hypotenuse* of a *right triangle.* The theorem states that the square of the *hypotenuse* is equal to the sum of the squares of the other two sides, or, in mathematical terms, $a^2+b^2=c^2$.

Rectangle: A four-sided surface in which all corners are right angles.

Right triangle: A three-sided surface with one angle that equals 90 degrees.

SI system: An abbreviation for Système International d'Unités, or International System of Units. The formal name of the *metric system.*

Screw: A *simple machine* consisting of an *inclined plane* wrapped around a *cylinder.*

Simple machine: A device that performs work in a single action. There are six types of simple machines: the *inclined plane,* the *lever,* the *pulley,* the *wedge,* the *screw,* and the *wheel and axle.*

Square: A *rectangle* in which all four sides are equal lengths. When used with another form of measurement, such as *square meter,* the term refers to a measure of area.

Square foot: In the *English system,* the basic measure of *area.* There are 144 square inches in one square foot.

Square meter: In the *metric system,* the basic measure of *area.* There are 10,000 square centimeters in a square meter.

Temperature: A measure of relative heat as measured by a scale.

Thermal expansion: The expansion of materials in all three dimensions when heated.

Thermometer: A tool used to measure *temperature.*

Volume: A measure of a total amount of space, measured in cubic units.

Wedge: A *simple machine* consisting of two inclined planes back-to-back.

Wheel and axle: A *simple machine* consisting of a circle attached to a central shaft that spins with the wheel.

Work: A measure of the force required to move an object a specified distance.

Conversion Tables

A. English to Metric

	To convert...	Into...	Multiply the English unit by...
LENGTH	Inches	Millimeters	25.40
	Feet	Centimeters	30.00
	Yards	Meters	0.90
	Miles	Kilometers	1.60
AREA	Square inches	Square centimeters	6.50
	Square feet	Square meters	0.09
	Square yards	Square meters	0.80
	Square miles	Square kilometers	2.60
	Acres	Hectares	0.40
MASS and WEIGHT	Fluid ounces	Grams	28.00
	Pounds	Kilograms	0.45
	Short tons	Megagrams	0.90
LIQUID MEASURE	Ounces	Milliliters	30.00
	Pints	Liters	0.47
	Quarts	Liters	0.95
	Gallons	Liters	3.80

B. Metric to English

	To convert...	Into...	Multiply the metric unit by...
LENGTH	Millimeters	Inches	0.040
	Centimeters	Feet	0.400
	Meters	Yards	1.100
	Kilometers	Miles	0.620
AREA	Square centimeters	Square inches	0.160
	Square meters	Square yards	1.200
	Square kilometers	Square miles	0.400
	Hectares	Acres	2.500
MASS and WEIGHT	Grams	Fluid ounces	0.035
	Kilograms	Pounds	2.200
	Megagrams	Short tons	1.100
LIQUID MEASURE	Milliliters	Ounces	0.034
	Liters	Pints	2.100
	Liters	Quarts	1.060
	Liters	Gallons	0.260

Area and Volume Formulas

Area Formulas		
Rectangle:	$A = lw$	(multiply length by width)
Right Triangle:	$A = \frac{1}{2}(bh)$	(one-half the product of the base and height)
Circle:	$A = \pi R^2$	(multiply pi [3.1416] by the square of the radius)

Volume Formulas		
Rectangular Prism:	$V = lwh$	(multiply length by width by height)
Right Triangular Prism:	$V = \frac{1}{2}(bwh)$	(multiply one-half the product of base by width by height)
Cylinder:	$V = (\pi R^2)h$	(multiply the area of the circle by the height)

Worksheet Illustration

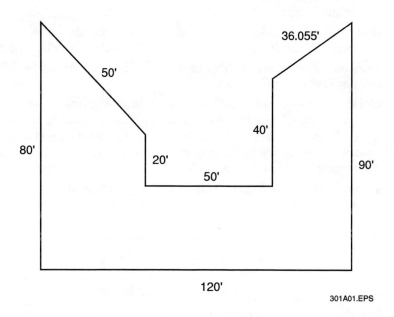

301A01.EPS

Additional Resources

This module is intended to present thorough resources for task training. The following reference works are suggested for further study. These are optional materials for continued education rather than for task training.

Code Check Plumbing: A Field Guide to Plumbing, 2000. Michael Casey, Douglas Hansen, and Redwood Kardon. Newton, CT: Taunton Press.

Math to Build On: A Book for Those Who Build, 1993. Johnny and Margaret Hamilton. Clinton, NC: Construction Trades Press.

Explosion Danger Lurks [Videotape], ca. 1955. North Andover, MA: Watts Regulator Company.

VHS video produced by Watts Regulator Company graphically demonstrates the dangers of excessive temperature rises in hot water heaters not having temperature and pressure relief valves. Covers the basic physics of temperature and pressure in domestic hot water systems.

Copies available from Watts Regulator Company, 815 Chestnut Street, North Andover, MA 01845. Call 1-800-617-3274.

Figure Credits

Expando Tools, Inc., and Sidu Manufacturing Company, Inc.	301F19
Kunkle Valve Company, Inc.	301F20
DeWalt Industrial Tool Company	301F26
Peninsula Fire Protection, Inc.	301F06, 301F08
Plumbing and Drainage Institute	301F17
Ridge Tool Company	301F25
Taylor Precision Products	301F12, 301F13, 301F14
Texas Instruments	301F04

References

Code Check Plumbing: A Field Guide to Plumbing, 2000. Michael Casey, Douglas Hansen, and Redwood Kardon. Newton, CT: Taunton Press.

Unit 52, "Water Pressure, Head, and Force" in *Mathematics for Plumbers and Pipefitters,* 5th ed., 1996. Bartholomew D'Arcangelo, Benedict D'Arcangelo, J. Russell Guest, and Lee Smith. Albany, NY: Delmar Publishers.

NCCER CRAFT TRAINING USER UPDATES

The NCCER makes every effort to keep these textbooks up-to-date and free of technical errors. We appreciate your help in this process. If you have an idea for improving this textbook, or if you find an error, a typographical mistake, or an inaccuracy in the NCCER's Craft Training textbooks, please write us, using this form or a photocopy. Be sure to include the exact module number, page number, a detailed description, and the correction, if applicable. Your input will be brought to the attention of the Technical Review Committee. Thank you for your assistance.

Instructors – If you found that additional materials were necessary in order to teach this module effectively, please let us know so that we may include them in the Equipment and Materials list in the Instructor's Guide.

Write: Curriculum Revision and Development Department
National Center for Construction Education and Research
P.O. Box 141104, Gainesville, FL 32614-1104

Fax: 352-334-0932

E-mail: curriculum@nccer.org

Craft _____ Module Name _____

Copyright Date _____ Module Number _____ Page Number(s) _____

Description _____
_____
_____
_____

(Optional) Correction _____
_____
_____

(Optional) Your Name and Address _____
_____
_____
_____

Codes

COURSE MAP

This course map shows all of the modules in the third level of the Plumbing curriculum. The suggested training order begins at the bottom and proceeds up. Skill levels increase as you advance on the course map. The local Training Program Sponsor may adjust the training order.

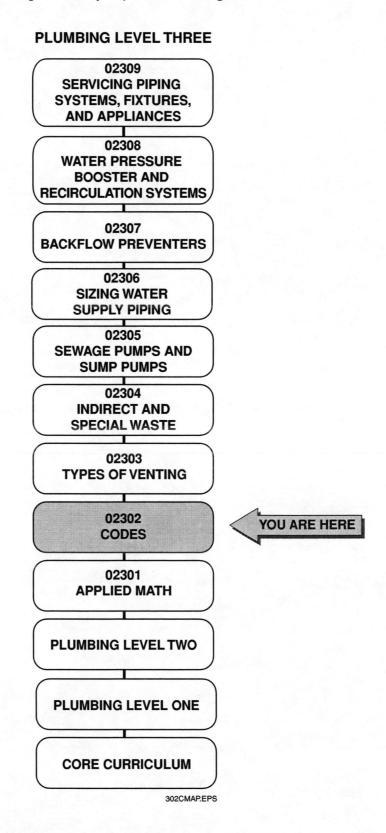

PLUMBING LEVEL THREE

02309
SERVICING PIPING SYSTEMS, FIXTURES, AND APPLIANCES

02308
WATER PRESSURE BOOSTER AND RECIRCULATION SYSTEMS

02307
BACKFLOW PREVENTERS

02306
SIZING WATER SUPPLY PIPING

02305
SEWAGE PUMPS AND SUMP PUMPS

02304
INDIRECT AND SPECIAL WASTE

02303
TYPES OF VENTING

02302
CODES ⟵ **YOU ARE HERE**

02301
APPLIED MATH

PLUMBING LEVEL TWO

PLUMBING LEVEL ONE

CORE CURRICULUM

302CMAP.EPS

MODULE 02302 CONTENTS

Figures

Codes

Objectives

When you have completed this module, you will be able to do the following:

1. Describe the model and local plumbing codes and their purposes.
2. Explain the procedure for modifying plumbing codes.
3. Demonstrate familiarity with the model code (if applicable) and local code used in your area.
4. Use the local plumbing code to find and cite references.

Prerequisites

Before you begin this module, it is recommended that you successfully complete the following: Core Curriculum; Plumbing Level One; Plumbing Level Two; Plumbing Level Three, Module 02301.

Required Trainee Materials

1. Appropriate personal protective equipment
2. Pencil and paper
3. Copy of your local code

1.0.0 ◆ INTRODUCTION

All construction work is governed by **codes.** A code is a legal document adopted in a jurisdiction that establishes the minimum acceptable standards, rules, and regulations for all materials, practices, and installations used in buildings and building systems. Codes are adopted to protect the health and safety of the public and property, ensuring that contractors perform their work according to recognized standards. In this module, you will learn about the different types of codes that affect plumbing installations, materials, and processes. You will also learn how codes are modified.

2.0.0 ◆ HISTORY OF CODES

Plumbing standards and practices observed in the United States today are based on those originally developed in Europe beginning in the 1500s. The predecessors of those standards can be traced to guidelines developed 500 years earlier to regulate safe building construction and public health in large urban settlements. Rudimentary regulations were incorporated into one of the earliest sets of codified laws. The Code of Hammurabi, written around 2100 B.C.E. for Hammurabi, King of Babylonia, declared that if a house collapsed, the builder should give his own house to the owner of the fallen house. If people died as a result of poor workmanship, the builder would lose his life, too. Modern building codes are not quite so drastic!

Not until the Renaissance in the sixteenth century did European civic leaders begin to enact laws requiring improved sanitation in the rapidly growing cities and trade centers. Beginning in 1519, for example, laws passed in various regions of France required that all houses should have indoor toilets or cesspools.

In the mid-nineteenth century, the British Parliament, faced with high mortality rates and substandard living conditions in urban areas throughout England, passed the 1848 Public Health Act. The first of its kind in scale and scope, the Public Health Act authorized a comprehensive program of proper water and sewer drainage, paved roads, and adequate supplies of clean water for drinking and bathing needs. It also gave power to local health boards to carry out the act's provisions.

Health laws benefited from advances in science. In 1864, French chemist Louis Pasteur developed his famous theory that infectious diseases were spread by microscopic organisms called germs. People had long suspected that contaminated water and unsanitary waste disposal were responsible for the rampant spread of disease. Now that people understood precisely how such things contributed to illness, they could develop sanitation technology to prevent such conditions from arising.

One of the first sanitation laws passed after Pasteur's discovery was the New York Metropolitan Health Law, enacted in 1866. At the time, the law was considered the most complete health legislation of its kind in the world. It served as a model for the health laws of other large cities. Over time, many of these health codes expanded to include examinations, training, and licensing for plumbers.

Early in the twentieth century, insurance companies in the United States called for the adoption of building codes to help reduce the excessive loss of lives due to fires in overcrowded inner cities. Over the next 40 years, code enforcement officials banded together in several regional professional organizations to develop and implement such codes, establishing the system that exists today.

3.0.0 ◆ MODEL CODES

There are an estimated 5,000 local codes in existence in the United States, and the potential total number of city, county, and state jurisdictions that could adopt codes is more than eight times that number—somewhere around 40,000. Obviously, not all of these codes have been created entirely from scratch. Many of them were developed from templates called **model codes.**

Model codes are comprehensive sets of general guidelines that establish and define acceptable plumbing practices and materials. They also list prohibited installations and clarify obvious plumbing hazards. Model codes do not have the force of law, but they serve as the basis for more detailed, legally binding local codes developed by individual jurisdictions, which can adopt and amend the model code in whole or in part. Because they offer general guidance, model codes tend to focus more on processes than on specifics.

Model codes are created by professional organizations with the expertise and resources to undertake such a large-scale task. These organizations work closely with government agencies, code administrators, and building industry professionals to develop, publish, and regularly revise the model codes. As well as plumbing

DID YOU KNOW?

Louis Pasteur, Plumber?

French chemist and biologist Louis Pasteur wanted to determine how microscopic organisms caused fermentation, which is the process whereby organic substances break down. Pasteur suspected that the agents of fermentation traveled through the air. To test his theory, he conducted a simple experiment involving a fermentable liquid poured into two flasks with S-shaped necks.

Pasteur first boiled the liquid to kill any microscopic organisms already in it. Then he poured the liquid into the two flasks. As the liquid cooled, condensation formed in the crooks of the flasks' necks, sealing the contents from the outside air. Pasteur then snapped the neck off one of the flasks, exposing the liquid to the air, and left them both in his lab overnight. When Pasteur examined the flasks the next day, he found that the liquid in the flask with the condensation seal showed no signs of fermentation. The liquid in the other flask, however, had grown cloudy from fermentation. Pasteur's experiment convinced colleagues and skeptics that his theory of airborne contamination was valid.

If the concept of using liquid in a curved tube to block exposure to the air sounds familiar to you, it should–Pasteur's experiment not only demonstrated the existence of airborne microscopic organisms, it also demonstrated the value of the humble, but indispensable, trap seal!

codes, model code organizations issue general standards for the following:

- Residential and commercial building construction
- Fire prevention
- Mechanical and electrical installations
- Energy conservation
- Property maintenance
- Sewage disposal
- Zoning

In general, professional code development organizations publish new editions of their codes every three years.

Most model code organizations are open to membership by professionals from the construction trades, architects and designers, regulation enforcement specialists, and industry representatives. These organizations have taken to the Internet to communicate with members and other interested parties. Web sites feature the latest news releases, provide detailed information about pending and approved code changes, and offer membership information. Many organizations also offer the option to purchase model codes and other publications online.

The world of model codes in the United States is a dynamic and rapidly changing one. In the past, model codes were oriented toward regional needs. Now they are gradually assuming a national and even international scope. Several code development organizations have recently combined their resources and expertise to develop new joint model codes or revise existing ones. As a result, some model code organizations now publish more than one set of model codes, while others have discontinued older codes in favor of newer ones.

Further complicating the issue, state and local jurisdictions may adopt new model codes for some construction standards but keep established codes for others. Some states develop their own codes without reference to a model code, and some use more than one model code. Because all of these issues pose challenges for plumbing professionals trying to keep up with new developments, a summary review of the various model code organizations and their activities will help clarify this complicated situation.

Contact information for each of the model code organizations discussed in this section as well as many others can be found in the *Appendix*.

3.1.0 Building Officials and Code Administrators International, Inc. (BOCA)

Established in 1915, the Building Officials and Code Administrators International, Inc. (BOCA), established the **BOCA National Plumbing Code**® (NPC), as well as national codes for building construction, mechanical installations, fire prevention, property maintenance, private sewage disposal, and energy conservation. Cities, counties, and states primarily in the eastern and midwestern United States have adopted the NPC (see *Figure 1*).

Along with the Southern Building Code Congress International, Inc., and the International Conference of Building Officials (ICBO), BOCA became a founding member of the International Code Council (ICC) in 1994. BOCA participates in the development, revision, and publication of the ICC's model codes; as a result, it is no longer updating the NPC.

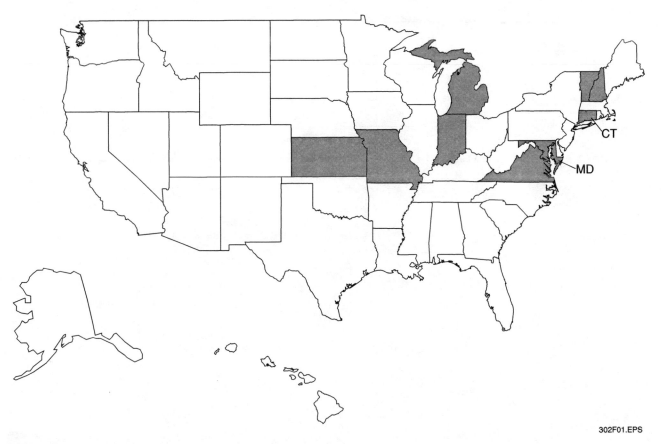

302F01.EPS

Figure 1 ◆ States adopting the BOCA National Plumbing Code® as of March 2001.

3.2.0 Southern Building Code Congress International, Inc. (SBCCI)

The Southern Building Code Congress International, Inc. (SBCCI), was founded in 1940. SBCCI published the first edition of its **Standard Plumbing Code™** (SPC) in 1955. Cities, counties, and states in the southern and southeastern United States (see *Figure 2*) have since adopted the SPC. SBCCI also issues model codes for mechanical installations, excavation and grading, gas, fire prevention, housing, swimming pools, existing building maintenance, building construction, and unsafe building abatement.

SBCCI was a co-founder of the International Code Conference (ICC) and is one of three model code organizations participating in the development, revision, and publication of the ICC's model codes. With the advent of the International Plumbing Code, SBCCI has stopped publishing the SPC.

3.3.0 International Code Council (ICC)

The International Code Council, Inc. (ICC) was founded in 1994 by three model code develop- ment organizations: the Building Officials and Code Administrators International, Inc. (BOCA), the International Conference of Building Officials (ICBO), and the Southern Building Code Congress International, Inc. (SBCCI). The ICC issued its first **International Plumbing Code®** (IPC) in 1995 and revised it in 2000. The ICC also publishes model codes, nicknamed *I-codes,* for building construction, fire prevention, residential buildings, mechanical installations, fuel gas, energy conservation, property maintenance, private sewage disposal, electrical installations, and zoning.

The ICC's goal is to develop a consistent and uniform set of international standards for the United States and eventually the world. The ICC sees a variety of benefits arising from global uniform standards:

- Construction professionals and code officials would be able to work with a single set of requirements throughout the world.
- Manufacturers would no longer have to design products for multiple standards.
- Professionals and manufacturers would be able to compete in worldwide markets.
- Education and certification programs would be standardized.

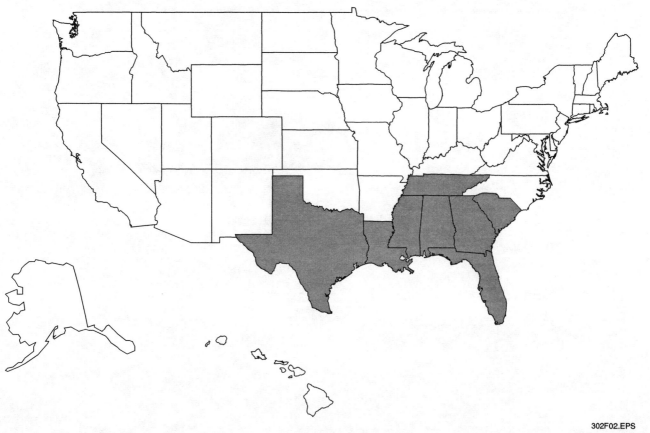

302F02.EPS

Figure 2 ◆ States adopting the SBCCI Standard Plumbing Code™ as of March 2001.

- Construction and enforcement standards would become consistent.
- Concerned parties would require only one forum to resolve code enforcement and regulation issues.

In the United States, cities and other local jurisdictions are gradually adopting the international codes for certain standards, while maintaining existing codes for others. Colorado, Delaware, the District of Columbia, Kansas, Ohio, Oklahoma, Rhode Island, Utah, and West Virginia have all adopted the IPC (see *Figure 3*).

3.4.0 International Association of Plumbing and Mechanical Officials (IAPMO)

The International Association of Plumbing and Mechanical Officials (IAPMO) was established in 1926. Originally formed as the Plumbing Inspectors Association of Los Angeles, the association changed its name to IAPMO in 1967. The organization has published the **Uniform Plumbing Code™** (UPC) since 1945. The UPC serves as a model state code in the western and midwestern United States and has been adopted by numerous other cities and counties throughout the country (see *Figure 4*).

Recently, IAPMO has formed partnerships with other model code organizations to revise existing guidelines. Currently, IAPMO is collaborating with the National Fire Protection Association (NFPA) to develop a new model building code, the plumbing guidelines of which are expected to be based on the UPC. In 1995, IAPMO, the Plumbing-Heating-Cooling Contractors—National Association (PHCC), and the Mechanical Contractors Association of America (MCAA) collaborated to revise the UPC. These same organizations also teamed with the American National Standards Institute (ANSI) to revise that organization's model plumbing code.

3.5.0 Plumbing-Heating-Cooling Contractors—National Association (PHCC)

One of the first organizations of its kind in the United States, the National Association of Master Plumbers (NAMP) was established in 1883. It later changed its name to the Plumbing-Heating-Cooling Contractors—National Association (PHCC) and began publishing the **National Standard Plumbing Code** (NSPC) in 1971.

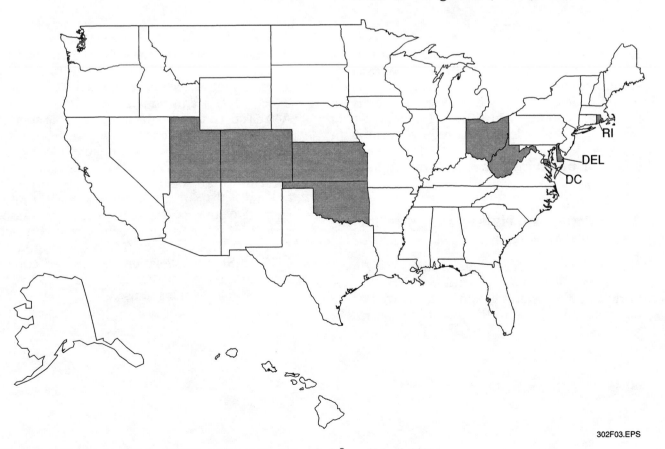

Figure 3 ◆ States adopting the ICC International Plumbing Code® as of March 2001.

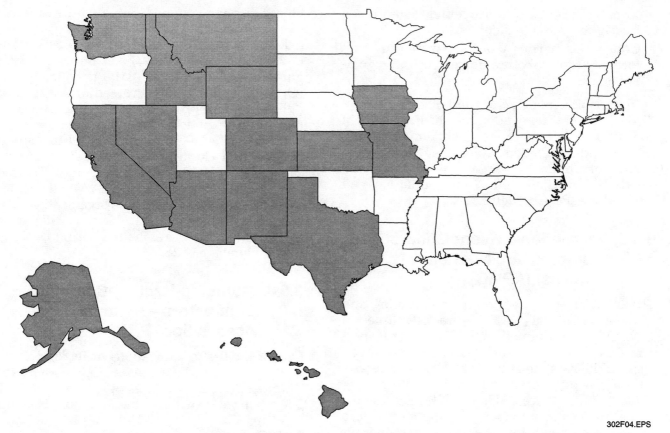

Figure 4 ◆ States adopting the IAPMO Uniform Plumbing Code™ as of March 2001.

In 1995, PHCC teamed with the International Association of Plumbing and Mechanical Officials (IAPMO) and the Mechanical Contractors Association of America (MCAA) to revise two model codes: IAPMO's UPC and the American National Standards Institute's (ANSI's) *Safety Requirements for Plumbing.* PHCC continues to revise and publish the NSPC, which serves as the basis for plumbing codes in three states (see *Figure 5*).

3.6.0 American National Standards Institute (ANSI)

Unlike other model code organizations, the American National Standards Institute (ANSI) is supported by the federal government as well as by many organizations in the private sector. Estab-

lished in 1918, ANSI's mission is to promote and enforce voluntary standards throughout the United States. In 1993, ANSI issued its latest model code for plumbing, which is designated ANSI A40 and titled **Safety Requirements for Plumbing.** ANSI A40 has been used as the model for one state code and in many jurisdictions and government installations throughout the United States (see *Figure 6*).

In 1995, an ANSI committee teamed up with the International Association of Plumbing and Mechanical Officials (IAPMO), the Plumbing-Heating-Cooling Contractors—National Association (PHCC), and the Mechanical Contractors Association of America (MCAA) to update the 1993 ANSI A40. Work on the revised code has stopped for the time being, and the 1993 version of ANSI A40 remains in use.

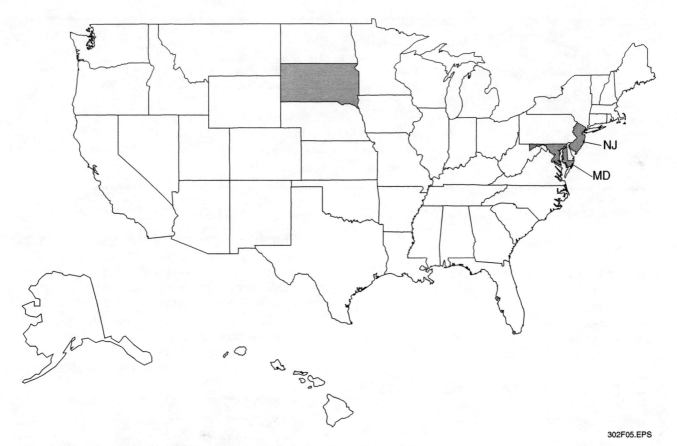

302F05.EPS

Figure 5 ◆ States adopting the PHCC National Standard Plumbing Code as of March 2001.

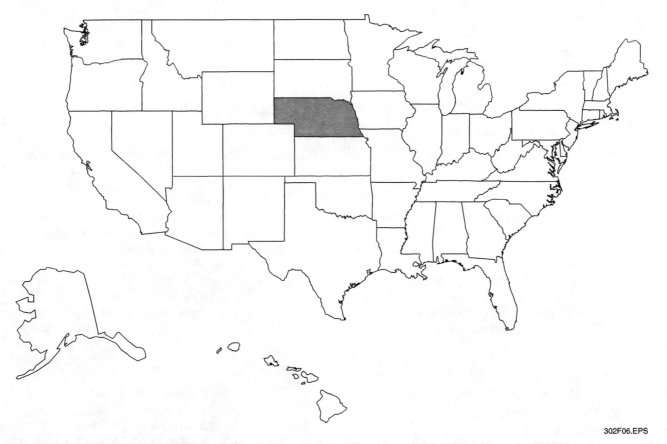

302F06.EPS

Figure 6 ◆ States adopting the ANSI A40 Safety Requirements for Plumbing model code as of March 2001.

Freeze Depths

Model codes can be used by jurisdictions anywhere in the country. Because of this fact, they do not provide detailed guidance for specific concerns that vary from region to region due to weather, climate, and humidity.

One issue for which model codes do not provide specific guidance is freeze depths. Wide variations in freeze depths occur throughout the country. Model codes simply state that plumbing should be protected from freezing. It is up to local officials to define these depths in the local codes.

Review Questions

Sections 2.0.0–3.0.0

1. Britain's Public Health Act was adopted in _____.
   a. 1519
   b. 1848
   c. 1866
   d. 1918

2. The New York Metropolitan Health Law was adopted in _____.
   a. 1848
   b. 1866
   c. 1915
   d. 1918

3. Model codes are published by _____.
   a. the federal government
   b. private plumbing consultants
   c. state code enforcement agencies
   d. professional organizations

4. The member organizations of the International Code Council are the Building Officials and Code Administrators International (BOCA), the International Conference of Building Officials (ICBO), and the _____.
   a. International Association of Plumbing and Mechanical Officials (IAPMO)
   b. Southern Building Code Conference International (SBCCI)
   c. American National Standards Institute (ANSI)
   d. Plumbing-Heating-Cooling Contractors—National Association (PHCC)

5. The International Association of Plumbing and Mechanical Officials (IAPMO) publishes the _____ Plumbing Code.
   a. National Standard
   b. International
   c. Uniform
   d. Standard

4.0.0 ◆ HOW CODES WORK

Plumbing codes are intended to reflect the current state of the science, technology, and practice of the trade. However, codes must change over time to include new materials and technologies and to reflect changing standards and practices. Both model and local codes change, though not always at the same time. Changes to model codes can drive changes to local codes and vice versa. The code change process is a constant cycle (see *Figure 7*).

DID YOU KNOW?

Low-Flush Toilets

In 1992 the U.S. Department of Energy mandated the use of low-flush toilets as a water conservation measure. These toilets use half the water of traditional toilets (1.6 gallons versus 3.5 gallons per flush), but, as a result, have less flushing power. These toilets are notorious for leaving residue in the bowl, clogging units, and requiring maintenance.

Drainage plumbing in older homes is designed to work with higher flow rates. As the flow of water comes through the pipe in the drainage system, it swirls in a vortex, helping to keep the pipe clean. When low-flush toilets are connected to pipes designed for older toilets, the water doesn't swirl; instead, it separates and doesn't clear the pipe. Thus, low-flush toilets cause clogs when there is not enough water flushing through the pipes. Additionally, when there is too much grade from the pipe to the ground, the same problem results—the water will simply run off and leave solids.

To combat these problems, it is important for plumbers to survey the water pressure in the home or office and to determine the appropriate toilet model. Available models include gravity tanks, which have a water pressure of 15 pounds per square inch (psi); pressure tanks, which require a minimum of 25 psi; and flushometers, which operate at 35 to 40 psi. Selecting the best toilet for the job will help alleviate problems in the future.

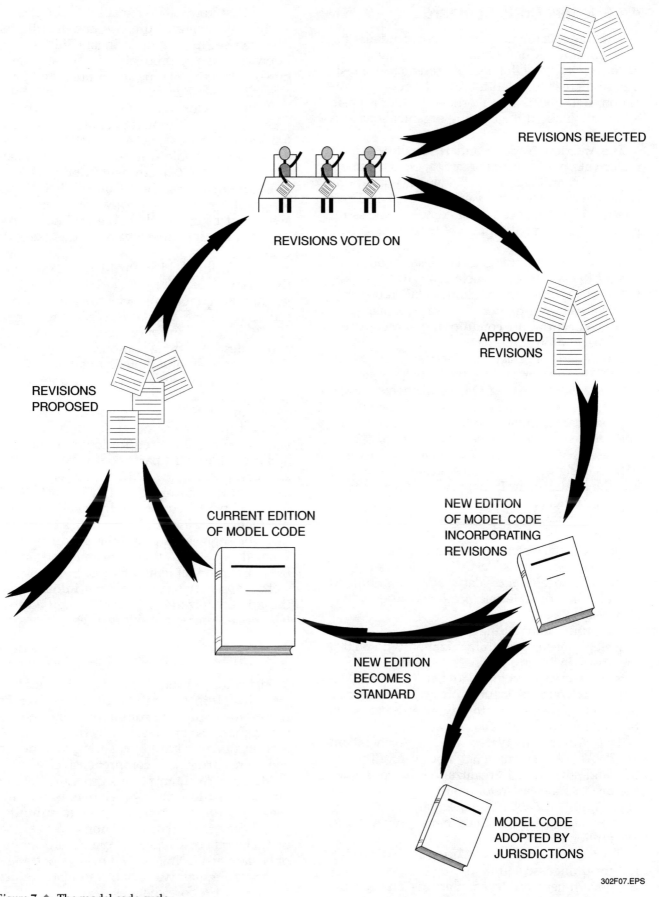

REVISIONS REJECTED

REVISIONS VOTED ON

APPROVED
REVISIONS

REVISIONS
PROPOSED

CURRENT EDITION
OF MODEL CODE

NEW EDITION
OF MODEL CODE
INCORPORATING
REVISIONS

NEW EDITION
BECOMES
STANDARD

MODEL CODE
ADOPTED BY
JURISDICTIONS

302F07.EPS

Figure 7 ◆ The model code cycle.

4.1.0 Model Code Standards

Several organizations oversee code standards for their specific industry. Standards organizations have already done the detailed research and testing, so there is no need for a model code organization to duplicate that work. Following is basic information about the existing organizations and the particular standards they set.

The American Nation Standards Institute (ANSI) is a private, not-for-profit organization that administers and coordinates the U.S. voluntary standardization and conformity assessment system. Their mission is to enhance global competitiveness by promoting and facilitating voluntary consensus standards.

The Air-Conditioning and Refrigeration Institute (ARI) is a trade association with 70 established standards and guidelines. ARI represents the manufacturers of more than 90 percent of U.S.-produced central air conditioning and commercial refrigeration equipment.

The American Society of Mechanical Engineers (ASME) is a not-for-profit, educational, and technical organization serving 125,000 members. ASME sets many industrial and manufacturing standards to promote and enhance the technical competency and professional well-being of its members.

The American Society of Sanitary Engineering (ASSE) designs programs to educate the industry and the public on the importance and necessity of safe and correct plumbing installation.

The American Society for Testing and Materials (ASTM) is a developer and provider of more than 10,000 voluntary consensus standards in 130 different technical fields. ASTM's mission is to promote public health and safety and to contribute to the reliability of materials, products, systems, and services.

The American Welding Society (AWS) is a multifaceted, not-for-profit organization with 50,000 members, including engineers, scientists, educators, researchers, welders, and inspectors. AWS supports welding education and technology development in order to advance the science, technology, and application of welding.

The American Water Works Association (AWWA) is an international, not-for-profit scientific and educational organization that promotes the health and welfare of quality drinking water and supply. AWWA has more than 50,000 members, including 4,000 utilities that supply water to 180 million people.

The Canadian Standards Association (CSA) is a not-for-profit association serving business, industry, government, and consumers in Canada. CSA develops standards that address and enhance public safety and preserve the environment.

The Cast Iron Soil Pipe Institute (CISPI) seeks to advance the manufacture, use, and distribution of cast iron soil pipe and fittings. CISPI strives to improve industry products, achieve standardization of cast iron soil pipe and fittings, and provide a continuous program of product testing, evaluation, and development.

Federal Specification (FS) establishes standards for the General Services Administration. FS standards are available from the Superintendent of Documents, U.S. Government Printing Office.

The International Code Council (ICC) is a not-for-profit organization dedicated to developing a single set of comprehensive and coordinated national model construction codes (see *Model Codes*).

The National Fire Protection Association (NFPA) promotes fire and electrical safety by providing and advocating scientifically based consensus codes and standards. NFPA has more than 75,000 members and more than 90 national trade and professional organizations.

The National Sanitation Foundation (NSF) is an independent, not-for-profit organization committed to public health, safety, and protection of the environment. NSF focuses on food, water, indoor air, and the environment; developing national standards; providing educational opportunities; and providing third-party conformity assessment services.

The Plumbing and Drainage Institute (PDI) includes manufacturers of engineered plumbing products in drainage. PDI promotes the advancement of engineered plumbing products through publicity, public relations, research, and standardization of product requirements.

The *Appendix* has the contact information for these 14 organizations, including their addresses, telephone numbers, and Web sites.

4.2.0 Revision of Model Codes

Model code revision is traditionally a public process, involving members of the construction trades, the regulatory community, and the industry. In short, the process may include anyone with a professional interest in ensuring that the model codes are current and comprehensive.

Model code organizations generally hold regular meetings to hear code change proposals. At these meetings, people are free to submit proposed changes for consideration and to argue for or against them. Change proposals are submitted on a standard form (see *Figure 8*). Any trade professional, regulatory official, or industry representative may propose changes.

For office use only. Date Rec'd. _____ Log No. _____ Proposal No. _____

PUBLIC PROPOSAL FORM

FOR PUBLIC PROPOSALS ON THE ICC CODES AND STANDARDS

(PLEASE SEE SUBMITTAL RULES OF PROCEDURES - ALL SUBMITTALS MUST BE IN COMPLIANCE WITH THESE PROCEDURES)

CLOSING DATE: All Proposals Must Be Received by the Announced Closing Date.

ICC

INTERNATIONAL CODE COUNCIL®

1) Indicate the format in which you would like to receive your Report of the Public Hearing (PRH), or Public Proposals Report (PPR):

   ☐ Paper   ☐ Electronic   ☐ Download

   (Note: A paper copy will not be sent to you if you have chosen to download the RPH or PPR from the ICC Web Site.)

2) **PLEASE TYPE OR PRINT CLEARLY: FORMS WILL BE RETURNED if they contain unreadable information.**

Name:		Date:	
Jurisdiction/Company:			
Submitted on Behalf of:			
Address:			
City:	State:	Zip +4:	
Phone:	Ext:	Fax:	
e-mail:			

3) ***Signature:** _____

   *I hereby grant the International Code Council the nonexclusive, royalty-free rights, including nonexclusive, royalty-free rights in copyright in this Proposal and I understand that I acquire no rights in any publication of International Code Council (ICC), in which this Proposal in this or another similar analogous form is used. I hereby attest that I have the authority and I am empowered to grant this copyright release.

4) Indicate appropriate ICC Code associated with this Public Proposal – Please use Acronym: _____
   (See instructions for list of Names and Acronyms for the I-Codes & I-Standards):

5) **Revision to:** ☐ Section _____ ☐ Table _____ ☐ Figure _____

6) **PROPOSAL** Revise as follows (check BOX and state proposed change):

   ☐ Revise as follows: ☐ Add new text as follows   ☐ Delete and substitute as follows: ☐ Delete without Substitution:

Show the proposed NEW or REVISED or DELETED TEXT in legislative format: ~~Line through text to be deleted.~~ Underline text to be added.

   ☐ PROPOSAL Continued (Attach additional sheets as necessary)

7) **SUPPORTING INFORMATION** (State purpose and reason, and provide substantiation to support proposed change):

   ☐ SUPPORTING INFORMATION Continued (Attach additional sheets as necessary)

PLEASE USE SEPARATE FORM FOR EACH PROPOSAL • SUBMITTAL IN ELECTRONIC FORMAT IS PREFERRED

(A 3 ½" disk saved in WordPerfect 6.0, 8.0 or Microsoft word 97 accompanying the hard copy would be appreciated)

e-mail: lbrown@intlcode.org   **Phone:** (703) 931-4533 x15   **Fax:** (703) 931-9128 or (703) 379-1546

Mail FORM and DISK to: ICC Program Manager, 5203 Leesburg Pike, Suite 600, Falls Church, VA 22041

ICC Proposal e-form   *Revised July 3, 2001*

302F08.TIF

Figure 8 ◆ A sample code change proposal form.

When all proposals have been submitted and all the arguments have been heard, the proposed changes are voted on. In most model code organizations, the vote is open to all public officials who represent cities, counties, and states that have adopted the model code. Some organizations allow a select committee of public officials to vote on the proposed code changes. Regardless of the method used, the approved changes are included in the next edition of the model code, which is usually published every three years.

4.3.0 Adoption of Model Codes

Model codes are comprehensive in scope, but they can offer only the minimum acceptable standards and the broadest guidelines. Therefore, the jurisdictions that adopt them are free to make changes and add amendments to accommodate their state and local regulations. Changes to model codes are typically determined by an appointed advisory commission of regulatory experts, the members of which are determined by law. The commission reviews the code that is current in the jurisdiction. They also review the most recent edition of various model codes unless the jurisdiction has developed its own.

The commission decides which model code to adopt and what, if any, changes or amendments are necessary. Adopted codes must be signed into law before they can take effect. Codes usually have to be submitted as an ordinance and approved by the regulatory officials of the jurisdiction, as well as by officials all the way up to the state government.

Adopted codes—building, mechanical, electrical, plumbing, fire protection—are reviewed on a regular basis, usually every three to five years. Reviews ensure that the jurisdiction's codes remain up-to-date and comprehensive. Otherwise, the constant need to amend codes with variances and exceptions for each new technology and process would result in a patchwork code and bureaucratic headaches. A local code may not change the requirements that are specified in the model code on which it is based.

The review process also gives regulators an opportunity to adopt a different model code from the one they have used in the past. Regulators may choose to adopt a new model code to bring the jurisdiction into closer alignment with neighboring jurisdictions, or the model code from which a local code was adopted may no longer be updated.

Authority for code administration, inspection, and enforcement varies from jurisdiction to jurisdiction. The responsibility may lie with a building inspector, the health department, the city engineer, or a commissioner's office. Take the time to become familiar with how codes are regulated in your area so that you will always know whom to call when you have questions or concerns about code compliance.

4.4.0 Typical Code Changes

Changes in model and local codes frequently result from developments in plumbing materials, the growing demand for energy and water conservation, and the availability of alternative energy sources. New materials are approved for inclusion in model and local codes only after they are proven consistently safe and reliable. For example, local codes increasingly accept the use of plastic pipe in plumbing installations, whereas years ago it was considered unsuitable for use in certain plumbing installations. In addition to the piping material itself, codes also approve the new fastening and joining methods that these pipes require.

Codes increasingly stress energy conservation measures, such as insulation for flow restrictors on hot and cold water outlets, pipes, and water heaters. Solar hot water installations are a popular alternative source of hot water in residential and commercial installations. They are addressed in special model codes, such as IAPMO's *Uniform Solar Energy Code* or the ICC's *Model Energy Code*. Be aware that model solar energy codes do not address specific technical issues, such as whether to use double-wall or single-wall heat exchangers or whether it is more appropriate to use toxic or nontoxic circulating fluids in solar water heater installations.

According to the ICC, the code development process includes the consideration of a code change proposal. The proposal must contain the purpose of the proposed change (for example, to update, clarify, revise), reasons for the change (in other words, justification), substantiation of the proposed change (including technical information), a bibliography, a completed copyright release form, and a cost impact statement. Before a public hearing, proposals are reviewed by the ICC office for compliance with the Rules of Proce-

dure. After discussion and review, the code change proposal goes through final action consideration, during which time the change is voted on. If approved, the change becomes part of the code.

5.0.0 ◆ TYPICAL CHAPTERS OF A MODEL CODE

With some minor variations, state and local codes are arranged in chapters that address proper procedures, correct sizing, installation requirements, and allowable materials in plumbing installations. Codes also point out practices and items that are prohibited. Here is a typical set of chapter headings for a code, based on the 2000 IPC:

- Administration—Found in every model code; includes the scope of the code, as well as information about plumbing officials, inspectors and inspections, records, permit fees, and enforcement
- Definition—Found in every model code
- General Regulations—Includes prohibited fittings, rodent prevention, excavation and backfill, and condensate disposal
- Fixtures, Faucets, and Fixture Fittings
- Water Heaters
- Water Supply and Distribution
- Sanitary Drainage
- Indirect/Special Waste
- Vents
- Interceptors and Separators
- Storm Drainage
- Special Piping and Storage Systems
- Referenced Standards—Lists all the regulations and standards cited in previous chapters
- Appendices—Includes sample plumbing permit fee schedule, rates of rainfall for various cities, guidelines for gray water recycling systems, standards for sizing a water piping system, notes on structural safety, and requirements for vacuum drainage systems

Be sure to study your local code carefully. Keep up with changes to the local code and stay informed about changes to the model code on which it is based. As a professional, you are always expected to ensure that your work meets or exceeds current regulatory standards and reflects the highest level of skill and workmanship.

Review Questions

Section 4.0.0–5.0.0

1. In most model code organizations, _____ has/ have the authority to approve changes to a model code.
   a. any member of the plumbing trade, industry, or regulatory body
   b. all members of the organization
   c. public officials
   d. members of an appointed commission

2. Model codes are changed to accommodate new materials and to _____.
   a. reflect industry preferences
   b. specify appropriate personal protective equipment
   c. reflect changing standards and practices
   d. reflect specific jurisdictional requirements

3. Adopted codes are typically reviewed every _____ years.
   a. 3 to 5
   b. 6 to 8
   c. 10
   d. 7

4. When a jurisdiction adopts a model code, code changes are typically determined by _____.
   a. industry representatives
   b. local regulatory officials
   c. the state governor
   d. a commission of appointed experts

5. The three most common catalysts for change in model and local codes are developments in plumbing materials, energy and water conservation, and _____.
   a. changes to building code standards
   b. alternative energy sources
   c. local geographical requirements
   d. proven safety and reliability

6.0.0 ◆ THE WORKSHEET

Use a copy of your local code to answer the following questions about requirements in your area for various types of plumbing installations. Be sure to cite the reference following your answer.

1. What are the duties and powers entrusted to your local code officials?

2. What work is specifically exempt from the requirement for a permit?

3. What are the requirements for connecting the drainage piping to offsets and bases of stacks?

4. How are vacuum system station receptacles to be installed?

5. What standard is referenced for the installation of nonflammable medical gas systems?

6. What are the minimum required air gaps for lavatories and fixtures with similarly sized openings?

7. How is the maximum water consumption flow rate determined for a public lavatory?

8. What is the maximum allowable amount of lead content in water supply pipe and fittings?

9. What are the requirements for joints between copper and copper-alloy tubing?

10. What are the size requirements for a shower compartment?

Summary

The purpose of a code is to provide safe and efficient plumbing installations for the public. Professional organizations develop model codes that serve as general guidelines for plumbing installations, materials, and practices. Jurisdictions can adopt and amend model codes to meet their specific requirements. Codes are regularly revised to reflect changing technologies and practices.

The plumber's work must be performed according to the standards and practices outlined in a code. The best way to understand a code is to take the time to read it thoroughly. Ask local officials and experienced plumbers to explain the similarities and differences between model, state, and local codes that apply to your area. Remember, codes are among the most important tools you will ever use on the job.

Trade Terms Introduced in This Module

BOCA National Plumbing Code® (NPC): A model plumbing code of the Building Officials and Code Administrators International, Inc. (BOCA). Adopted primarily in the eastern and midwestern United States, it is no longer updated.

Code: A legal document enacted to protect the public and property, establishing the minimum standards for materials, practices, and installations.

International Plumbing Code® (IPC): The model plumbing code of the International Code Council (ICC). It was first issued in 1995 and is currently adopted in nine states and several local jurisdictions.

Model code: A set of comprehensive, general guidelines that establish and define acceptable plumbing practices and materials and list prohibited installations.

National Standard Plumbing Code (NSPC): The model plumbing code of the Plumbing-Heating-Cooling Contractors—National Association (PHCC). It was first published in 1971 and adopted as the state code by three states and several local jurisdictions.

Safety Requirements for Plumbing: A model plumbing code from the American National Standards Institute (ANSI). It was last updated in 1993 and is currently adopted in one state, several local jurisdictions, and many federal installations.

Standard Plumbing Code™ (SPC): A model plumbing code of the Southern Building Code Congress International. Introduced in 1955 and adopted by cities, counties, and states in the southern and southeastern United States, it is no longer updated.

Uniform Plumbing Code™ (UPC): The model code of the International Association of Plumbing and Mechanical Officials (IAPMO). It has been published since 1945 and was adopted by states in the western and midwestern United States.

Model Code Organizations

Air-Conditioning and Refrigeration Institute
(ARI)
4301 N. Fairfax Drive, Suite 425
Arlington, VA 22203
(703) 524-8800
www.ari.org

American National Standards Institute (ANSI)
1819 L Street, NW, 6th Floor
Washington, DC 20036
(202) 293-8020
www.ansi.org

American Society of Mechanical Engineers
(ASME)
Three Park Avenue
New York, NY 10016-5990
(212) 591-8500
www.asme.org

American Society of Sanitary Engineering
(ASSE)
901 Canterbury, Suite A
Westlake, OH 44145
(440) 835-3040
www.asse-plumbing.org

American Society for Testing and Materials
(ASTM)
100 Barr Harbor Drive
West Conshohocken, PA 19428-2959
(610) 832-9585
www.astm.org

American Water Works Association (AWWA)
6666 West Quincy Avenue
Denver, CO 80235
(303) 794-7711
www.awwa.org

American Welding Society (AWS)
550 N.W. LeJeune Road
Miami, FL 33126
(305) 443-9353
www.aws.org

Building Officials and Code Administrators
International, Inc. (BOCA)
4051 West Flossmoor Road
Country Club Hills, IL 60478-5795
(800) 214-4321
www.bocai.org

Canadian Standards Association (CSA)
178 Rexdale Boulevard
Toronto, Ontario, Canada M9W 1R3
(416) 747-4044
www.csa.ca

Cast Iron Soil Pipe Institute (CISPI)
5959 Shallowford Road, Suite 419
Chattanooga, TN 37421
(423) 400-5842
www.cispi.org

Federal Specifications (FS)
General Services Administration
7th and D Streets
407 E. L'Enfant Plaza, SW, Suite 8100
Washington, DC 20024-2124
(800) 688-9889
www.gsa.gov

International Association of Plumbing and
Mechanical Officials (IAPMO)
20001 E Walnut Drive, South
Walnut, CA 91789-2825
(909) 595-8449
www.iapmo.org

International Code Council, Inc. (ICC)
5203 Leesburg Pike, Suite 600
Falls Church, VA 22041-3405
(703) 931-4533
www.intlcode.org

National Fire Protection Association (NFPA)
1 Batterymarch Park
Quincy, MA 02269-9101
(617) 770-3000
www.nfpa.org

National Sanitation Foundation (NSF)
789 N. Dixboro Road
P.O. Box 130140
Ann Arbor, MI 48113-0140
(734) 769-8010
www.nsf.org

Plumbing and Drainage Institute (PDI)
45 Bristol Drive
South Easton, MA 02375
(800) 589-8956
www.pdionline.org

Plumbing-Heating-Cooling Contractors –
National Association (PHCC)
180 South Washington Street
P. O. Box 6808
Falls Church, VA 22040
(800) 533-7694
www.phccweb.org

Southern Building Code Congress
International, Inc. (SBCCI)
900 Montclair Road
Birmingham, AL 35213-1206
(205) 591-1853
www.sbcci.org

Additional Resources

This module is intended to present thorough resources for task training. The following reference works are suggested for further study. These are optional materials for continued education rather than for task training.

Code Check: A Field Guide to Building a Safe House, 2000. Redwood Kardon, Michael Casey, and Douglas Hansen. Newtown, CT: Taunton Press.

Code Check Plumbing: A Field Guide to Plumbing, 2000. Redwood Kardon, Michael Casey, and Douglas Hansen. Newtown, CT: Taunton Press.

The Engineering Resources Code Finder for Building and Construction, 2001. Dennis Phinney. Anaheim, CA: BNI Building News.

References

1997 International Plumbing Code, 1991. Falls Church, VA: International Code Council, Inc.

Code Check Plumbing: A Field Guide to Plumbing, 2001. Redwood Kardon, Michael Casey, and Douglas Hansen. Newtown, CT: Taunton Press.

Modern Plumbing, 1997. E. Keith Blankenbaker. Tinley Park, IL: Goodheart-Willcox Company.

Acknowledgments

International Code Council, Inc. (for use of the Table of Contents for the *2000 International Plumbing Code,* 2001. Falls Church, VA: International Code Council, Inc.), 5203 Leesburg Pike, Suite 600, Falls Church, VA 22041-3401, (703) 931-4533.

Studor, Inc. (for current information about state model code adoptions included in their product catalog ["Plumbing Code Adoptions," in *Studor Air Admittance Valves,* Product Catalog, April 2000 edition. Dunedin, FL: Studor Inc.]).

The NCCER makes every effort to keep these textbooks up-to-date and free of technical errors. We appreciate your help in this process. If you have an idea for improving this textbook, or if you find an error, a typographical mistake, or an inaccuracy in the NCCER's Craft Training textbooks, please write us, using this form or a photocopy. Be sure to include the exact module number, page number, a detailed description, and the correction, if applicable. Your input will be brought to the attention of the Technical Review Committee. Thank you for your assistance.

Instructors – If you found that additional materials were necessary in order to teach this module effectively, please let us know so that we may include them in the Equipment and Materials list in the Instructor's Guide.

Write: Curriculum Revision and Development Department
National Center for Construction Education and Research
P.O. Box 141104, Gainesville, FL 32614-1104

Fax: 352-334-0932

E-mail: curriculum@nccer.org

Craft   Module Name

Copyright Date   Module Number   Page Number(s)

Description

(Optional) Correction

(Optional) Your Name and Address

Types of Venting

COURSE MAP

This course map shows all of the modules in the third level of the Plumbing curriculum. The suggested training order begins at the bottom and proceeds up. Skill levels increase as you advance on the course map. The local Training Program Sponsor may adjust the training order.

PLUMBING LEVEL THREE

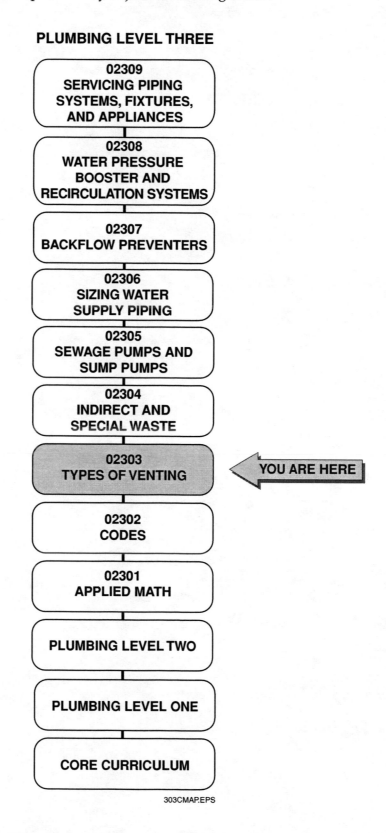

02309
SERVICING PIPING
SYSTEMS, FIXTURES,
AND APPLIANCES

02308
WATER PRESSURE
BOOSTER AND
RECIRCULATION SYSTEMS

02307
BACKFLOW PREVENTERS

02306
SIZING WATER
SUPPLY PIPING

02305
SEWAGE PUMPS AND
SUMP PUMPS

02304
INDIRECT AND
SPECIAL WASTE

02303
TYPES OF VENTING

YOU ARE HERE

02302
CODES

02301
APPLIED MATH

PLUMBING LEVEL TWO

PLUMBING LEVEL ONE

CORE CURRICULUM

303CMAP.EPS

MODULE 02303 CONTENTS

Figures

Types of Venting

Objectives

When you have completed this module, you will be able to do the following tasks in accordance with local codes:

1. Demonstrate an understanding of the scientific principles of venting.
2. Design vent systems according to local code requirements.
3. Sketch the different types of vents.
4. Construct given vent configurations.
5. Install the different types of vents correctly.

Prerequisites

Before you begin this module, it is recommended that you successfully complete the following: Core Curriculum; Plumbing Level One; Plumbing Level Two; Plumbing Level Three, Modules 02301 and 02302.

Required Trainee Materials

1. Appropriate personal protective equipment
2. Pencil and paper
3. Copy of your local code

1.0.0 ◆ INTRODUCTION

Vents allow air to enter and exit a drain, waste, and vent (DWV) stack. Vents work in conjunction with drains to enable wastes to flow from fixtures into the waste stack, where they are carried into the sewer. Without vents, plumbing systems would not work. Odors and contaminants would build up, and people would become sick. With proper venting, wastes are carried away efficiently and effectively.

In *Plumbing Level Two*, you were introduced to vents as components of DWV systems. In this module you will learn how vents operate. You will also review the different types of vents that can be installed in a DWV system. Many types of vents can be grouped together into broad categories. This will make it easier for you to remember what they are called and how they are used.

2.0.0 ◆ HOW VENTS OPERATE

Plumbers need to understand how DWV systems operate. This allows plumbers to design and install DWV systems that function well. Fortunately, the principles are easy to understand.

DWV systems work by maintaining equalized pressure throughout the drainage system. Air is used to equalize pressure that is exerted when waste water drains from a fixture into the system.

ON THE
·LEVEL·

Purpose of Vents

In plumbing, vents, despite their different uses, are designed for the same purpose—to allow air to circulate in a DWV system. Without air circulation, a DWV system could not work.

You can visualize how a vent works by observing the behavior of water in a straw. If you fill a straw with water and put your thumb over the top, the water remains in the straw. As soon as you remove your thumb, the water drains out. This is because water cannot flow out of the straw unless air can flow into the straw at the same rate, filling the space left by the draining water. In other words, by replacing the water in the straw with an equivalent amount of air, an equal pressure is maintained inside the straw.

To achieve the same effect in a DWV installation, plumbers install vents. Vents allow air to enter the system when wastewater flows out of the drain pipe. They ventilate the drainage system to the outside air and keep the pipes from clogging. Air pressure, therefore, is important for creating and maintaining proper flow in drainage pipes. Also, normal air pressure throughout the DWV system helps maintain the water seal in fixture traps. System pressure should be equal to air pressure.

Too much or too little air pressure in the vent pipes can cause the DWV system to malfunction. Low pressure in the waste piping (also called a *partial vacuum* or *negative pressure*) can suck the water seal out of a trap and into the drain. This is called **self-siphonage** (see *Figure 1*). When the discharge from another drain creates lower pressure in the system, it can draw a trap seal into the waste pipe (see *Figure 2*). This is called **indirect or momentum siphonage**. Properly placed vents prevent both types of siphonage. To further protect against indirect siphonage, ensure that the stack is properly sized.

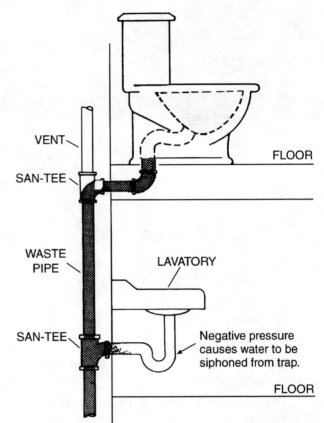

NOTE: Installation is for the purpose of demonstration only.

303F02.TIF

Figure 2 ◆ Indirect or momentum siphonage.

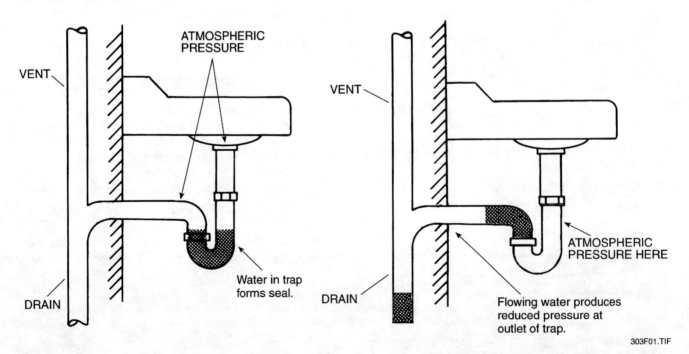

Figure 1 ◆ Self-siphonage.

303F01.TIF

Likewise, excess pressure in vent piping can have the reverse effect, blowing the trap seal out through the fixture. Air in the stack that is compressed by the weight of water above can force the water out of a trap as the slug of air passes a fixture. This phenomenon, called **back pressure**, has a tendency to happen in taller buildings (see *Figure 3*). Use a vent near the fixture traps or at the point where the piping changes direction to prevent back pressure. In some cases, downdrafts of air entering a vent on a roof can create a condition similar to back pressure. Install roof vents away from roof ridges and valleys that stir up downdrafts.

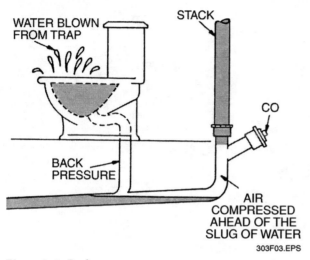

Figure 3 ◆ Back pressure.

3.0.0 ◆ DESIGNING A VENT INSTALLATION

A vent system must provide adequate venting for all the fixtures in a building. Each building has its own special requirements, which are dictated by:

• Location
• Use
• Level of occupancy

Vent systems must also be specially designed. However, most vent systems, regardless of where and how they are installed, have the same basic components. Plus, vents can be installed using procedures that are already familiar to you.

DID YOU KNOW?
Venting Medical Gas

Medical gas systems affect the treatment and care of hospital patients. Venting medical gas is a complicated task that requires additional training. To be able to complete this work, you must be at least a journeyman and have a medical gas certification. If you are not properly trained and certified, do not attempt to do this work. Seek the assistance of a trained professional.

DID YOU KNOW?
The Discovery of Air Pressure

In 15th century Italy, the Grand Duke of Tuscany ordered his engineers to build a giant pump. The pump was designed to raise water more than 40 feet above the ground. It worked by drawing water into a tall column by creating a partial vacuum. However, the engineers found that the pump could not raise water higher than 32 feet. The Grand Duke asked Evangelista Torricelli to solve the mystery. Torricelli (1608–1647) was a famous mathematician and a physicist.

Torricelli discovered that the pump worked fine; something else was preventing the water from reaching a higher level. To find the answer to the problem, he conducted a simple experiment. He filled a three-foot long test tube with liquid mercury. Then he quickly turned the tube upside down into a bowl full of more mercury, making sure the mouth of the tube was below the surface of the mercury in the bowl.

Some of the mercury in the tube flowed into the bowl, but most stayed inside the upside-down tube. Between the column of mercury in the tube and the bottom—now the top—of the tube, there was a gap, a vacuum like the one created in the duke's pump.

Torricelli reasoned that the air in the atmosphere was pushing down on the mercury in the bowl, forcing some of the mercury to stay in the tube. We call Torricelli's pushing force *air pressure*. The same thing was happening to the duke's pump. The water could only rise to 32 feet because the weight of the surrounding air could not push it any higher.

We now know that air pressure at sea level exerts a pressure of approximately 14.7 pounds per square inch. In other words, at any given moment you are being gently squeezed on all sides by the weight of the air around you. Air pressure affects things that most people may never think about: how long it takes water to boil, how birds fly, and how a pump works.

3.1.0 Components of a Vent System

The relationship between the vents in a DWV system is illustrated in *Figure 4*. The main soil and waste stack runs between the building drain and the highest horizontal drain in the system. Above the highest horizontal drain is the **stack vent**, which is an extension of the main soil and waste stack. The stack vent allows air to enter and exit the plumbing system through the roof. The stack vent also serves as the **vent terminal** for other vent pipes.

The central vent in a building is called the **vent stack**. The vent stack permits air circulation between the drainage system and the outside air. Other vents may be connected to it. The vent stack, which is also called the *main vent*, runs vertically. It is usually located within a few feet of the main soil and waste stack. The vent stack begins at the base of the main stack and continues until it connects with the stack vent. It may also extend through the roof on an independent path. Branches connect one or more individual fixture vents to the vent stack.

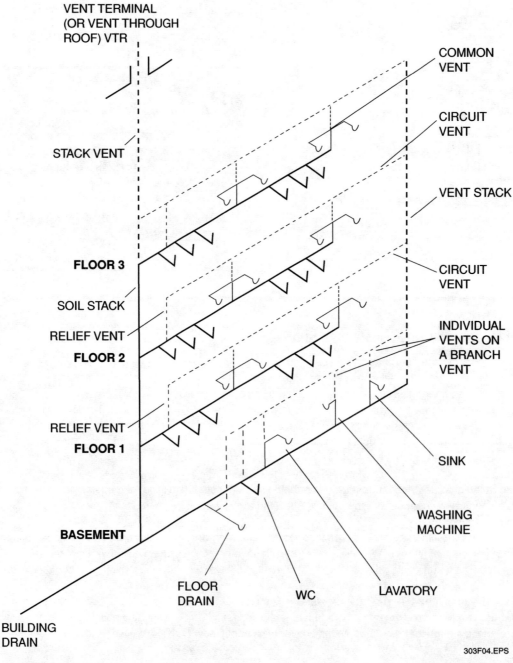

Figure 4 ◆ Vents in a DWV system.

303F04.EPS

Vents that run horizontally to a drain are called **flat vents**. Flat vents should not be used in floor-mounted fixtures. They can cause waste to back up into the vent and block the fixture drain (see *Figure 5*).

The vent terminal is the point at which, appropriately, a vent terminates. Fixture vents terminate at the point where they connect to the vent stack or directly to the stack vent. Stack vents and some vent stacks terminate above the roof. Local codes govern the location, size, and type of roof vents (see *Figure 6*). Follow the code closely when selecting and installing a roof vent terminal. When installing a roof vent, consider the following issues:

- The length of extension above the roof line
- The type of roof top
- The location of other structures on the roof
- The need for protection from rain, snow, excessive wind, and freezing

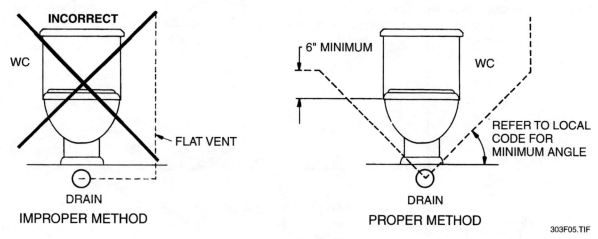

Figure 5 ◆ Improper and proper vent installation for a floor-mounted fixture.

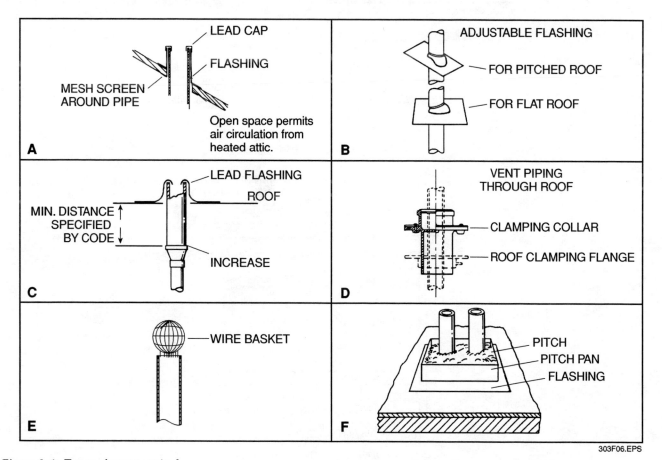

Figure 6 ◆ Types of vent terminals.

3.2.0 Vent Grades

Grade is the slope or fall of a line of pipe in reference to a horizontal plane. In a DWV system, grade allows solid and liquid wastes to flow out under the force of gravity. You learned how to calculate and measure grade in a DWV system in *Plumbing Level Two*.

Horizontal vent pipes also need to be installed with proper grade. This will allow condensation to drain without blocking the vent pipe. Your local plumbing code should specify the degree of slope required for a horizontal vent. Most codes require that the vent be taken off above the soil stack centerline. Then it must rise at an angle, typically not

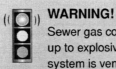

less than 45 degrees from horizontal, to a point 6 inches above the fixture's flood level rim (see *Figure 7*). That way, if the drain pipe becomes blocked, solids cannot enter and potentially clog the vent instead.

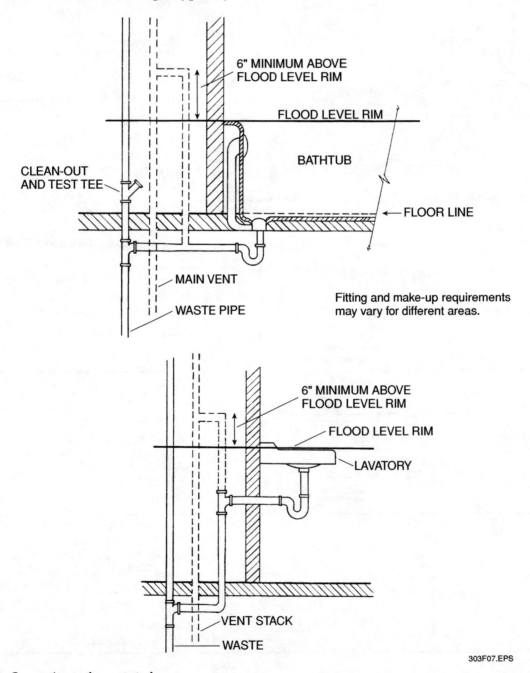

Figure 7 ◆ Connection to the vent stack.

The maximum grade between the trap weir and the vent pipe opening should not exceed one pipe diameter (see *Figure 8*). This grade is called the **hydraulic gradient**. The gradient determines the distance from the weir to the vent. If the vent is placed too far from the fixture trap, the vent opening will end up below the weir. This could cause self-siphoning. Most codes include tables that specify the maximum distance the weir can be positioned from the vent based on pipe fall and diameter.

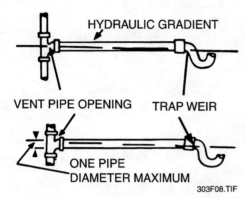

Figure 8 ◆ Determining the hydraulic gradient.

A vent opening that is located within two pipe diameters of a trap is called a **crown vent** (see *Figure 9*). Crown vents are prohibited by code. Relocate such vents further away from the trap.

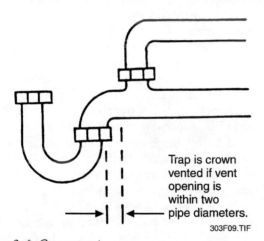

Trap is crown vented if vent opening is within two pipe diameters.

Figure 9 ◆ Crown vent.

Review Questions

Sections 2.0.0–3.0.0

1. Vents allow _____ to equalize the pressure in a DWV system.
   a. waste
   b. water
   c. air
   d. solids

2. Self-siphonage occurs when _____ in the waste piping sucks a water seal into a drain.
   a. back pressure
   b. low pressure
   c. high pressure
   d. excess pressure

3. Back pressure can be prevented by installing a vent in one of two places: near the fixture traps or _____.
   a. at the connection to the stack vent
   b. at the connection to the vent stack
   c. where the stack terminates
   d. where the piping changes direction

4. The _____ is located immediately above the highest horizontal drain in a DWV system.
   a. stack vent
   b. vent stack
   c. vent terminal
   d. crown vent

5. The hydraulic gradient is the maximum grade between a trap weir and the _____.
   a. nearest flat vent
   b. vent pipe opening
   c. drain pipe opening
   d. nearest horizontal vent

4.0.0 ◆ TYPES OF VENTS

Every venting system must be custom designed to provide the most efficient venting for a building, taking into account a host of different factors:

- The building's function
- The building's physical layout
- The building's location
- Applicable local codes

There are many different types of vents from which to choose. That way, plumbers have the flexibility to create the best DWV system possible. Each vent is constructed to serve a particular purpose.

Vents are used in a variety of combinations with each other. Many vents have more than one name, depending on the code being used. To make it easier to remember all the different types and names of vents, it will help to begin by grouping them together into a few broad categories.

4.1.0 Individual Vents

A vent that runs from a single fixture directly to the vent stack is called an **individual vent**. Plumbers install, or **re-vent**, individual vents to provide correct air pressure at each trap. Because of this, they are perhaps the most effective means of venting a fixture. A **continuous vent** is a type of individual vent. It is a vertical extension of the drain to which it is connected. Install continuous vents far enough away from fixture traps to avoid crown venting. Connect continuous vents to the stack at least 6 inches above the fixture's flood level rim or overflow line.

A **back vent**, as its name suggests, is an individual vent installed at the back of a fixture. Use back vents to connect the fixture drain pipe to the vent stack, or terminate back vents in the open air. In most cases, back vents should be connected to the drain pipe as close to the trap as possible. Back vents may incorporate a continuous vent (refer to *Figure 10*). Appropriate connection points, slope, and pipe sizes for back vents are spelled out in the building plans or your local code. A **branch vent** is any vent that connects a single fixture or a group of fixtures to a vent stack.

4.2.0 Common Vents

Use a **common vent** where two similar fixtures are installed with all three of the following characteristics:

- The fixtures are back-to-back.
- The fixtures are on opposite sides of a wall.
- The fixtures are at the same height.

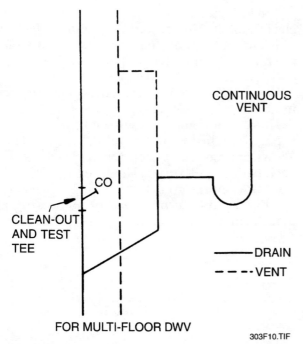

Figure 10 ◆ Back vent incorporating a continuous vent.

Common vents are also called unit vents. In a common vent, both fixture traps are connected to the drain with a short pattern sanitary cross (see *Figure 11*). A sanitary cross joins two back-to-back fixtures to a single drain pipe. The two fixtures share a single vent. The vent extends above the drain from the top connection of the sanitary cross. Install common vents in apartments or hotels where the design calls for the use of back-to-back fixtures and shared vents.

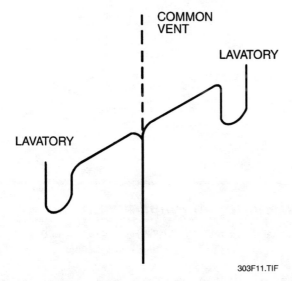

Figure 11 ◆ Common or unit vent.

4.3.0 Battery Vents

Many commercial buildings have large numbers of fixtures. It may be impractical or even unnecessary to provide each fixture with an individual vent. Horizontal vents connect a series, or *battery*, of fixtures to the vent stack (see *Figure 12*). **Circuit vents** and **loop vents** connect horizontal vents to the vent stack or stack vent. Select the proper type of battery vent for the application, and maintain the same pipe sizes throughout the vent. Local codes vary in their requirements for battery vents. Refer to your local code before venting a fixture battery.

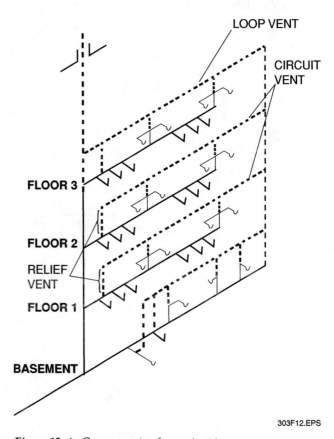

Figure 12 ◆ Components of a vent system.

A vent that connects a battery to the vent stack is called a *circuit vent*. In this arrangement, the circuit vent is connected to the drain before the first fixture and the last fixture and to the vent stack.

The loop vent is a special type of circuit vent. Install loop vents from the fixture group to the stack vent. Some codes require that fixture batteries on the top floors of a building be vented with loop vents.

4.4.0 Wet Vents

Vent piping that also carries liquid waste is a **wet vent**. It eliminates the need to install separate waste and vent stacks. Because of this, wet vents are a popular option for venting residential bathroom fixtures. Install wet vents either vertically or horizontally (see *Figure 13*). Only use wet vents with fixtures that have a comparatively low rate of flow, such as:

- Bathtubs
- Showers
- Bidets
- Lavatories
- Drinking fountains

When installing wet vents on toilets or water closets, follow the local code carefully. Make sure that the wet-vented fixture traps are capable of resealing. Maintain the proper hydraulic gradient and correct distances to ensure resealing.

Wet vents are considered acceptable by most model codes. Model codes cite the maximum vertical and horizontal run allowed for wet-vented pipe. The Uniform Plumbing Code™ (UPC) allows the use of wet vents for single and back-to-back bathroom fixtures. However, the UPC does not allow horizontal wet venting. Check your local code to determine the proper sizing and placement of a wet vent.

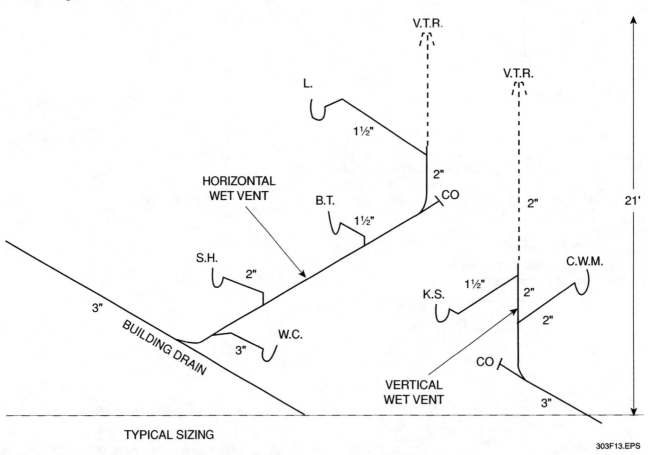

Figure 13 ◆ Wet vents.

Wet Vents and the International Plumbing Code

The latest edition of the International Plumbing Code® (IPC) was issued in 2000. Its guidelines for horizontal and vertical wet vents are much less restrictive than previous model codes.

For example, the IPC does not specify the order that bathroom fixtures should be placed. Other model codes require the water closet to be the last fixture in a bathroom wet vent. Also, the IPC permits a slope of ⅛ inch per foot for a 3-inch wet vent pipe. If your local code is based on the IPC, review it closely to learn the new requirements for wet vents.

4.5.0 Air Admittance Vents

Air admittance vents ventilate the DWV stack using the building's own air (see *Figure 14*). Air admittance vents are also called *air admittance valves*. They are an alternative to vent stacks that penetrate building roofs. Air admittance vents operate like valves by allowing a controlled one-way flow of air into the drainage system. Reduced pressure causes the vent to open and allows air to enter the drain pipes. When the pressure is equalized, the vent closes by gravity. If back pressure occurs, the vent seals tightly to prevent sewer gases from escaping.

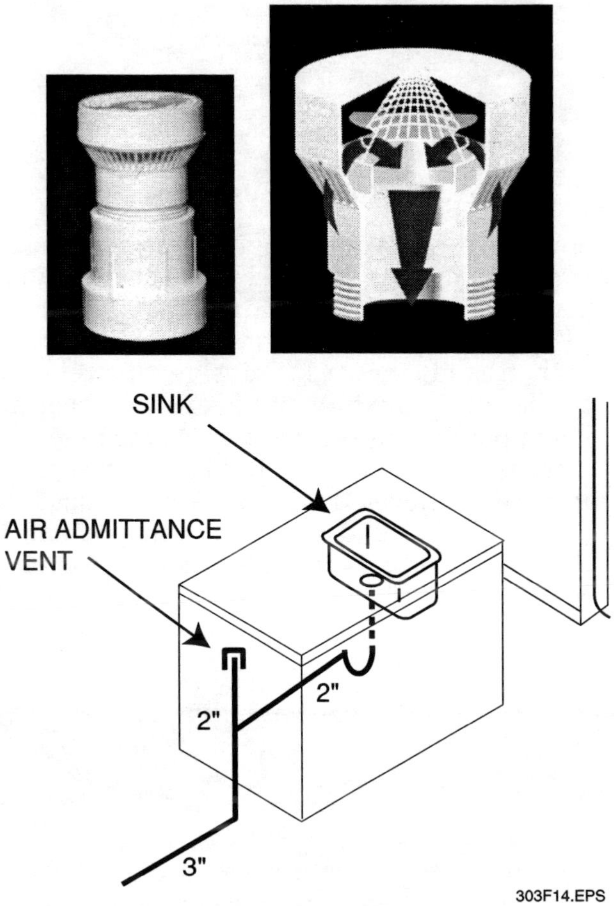

![Air admittance vent figure]

Figure 14 ◆ Air admittance vent.

Air admittance vents come in two sizes. The large size vents entire systems. The smaller size vents individual fixtures. Most model codes permit air admittance valves in residential buildings. The International Plumbing Code® (IPC), for example, allows air admittance valves on branch vents and individual vents. In the future, it may permit their use in stack vents as well. However, not all local codes accept air admittance valves. Check your local code for the proper sizing and placement of air admittance vents.

Use air admittance vents to vent fixtures in the center of a room, such as kitchen islands and wet bars. Codes cite maximum permitted distances between an island fixture and a main vent. Other options are to use a loop vent or a **combination waste and vent system** under the fixture (see *Figure 15*). This is a pipe that acts as a vent and also drains wastewater. If a combination waste and vent is used, ensure that the drain pipe is one size larger than the trap. The options for venting island fixtures depend on local codes.

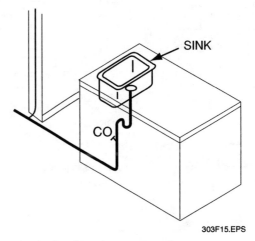

Figure 15 ◆ Combination waste and vent system.

4.6.0 Relief Vents

A **relief vent** allows excess air pressure in the drainage system to escape. This prevents back pressure from blowing out the seals in the fixture traps (see *Figure 16*). Use relief vents to increase circulation between the drain and vent systems. They can also serve as an auxiliary vent. A relief vent that is connected between the soil stack and the vent stack is called a **yoke vent**. Relief vents can also be connected between the soil stack and the fixture vents. In either case, connect the relief vent to the soil stack in a **branch interval**, which is the space between two branches entering a main stack. The space between branch intervals is usually story height but cannot be less than 8 feet.

Relief vents are very important in tall buildings. The weight of the water column in the stack can compress air trapped below it. Install relief vents on the floors of a high-rise building as called for in the design. A common practice is to install offset and yoke vents every 10 floors. Review your local code for specific details regarding the proper installation of relief vents.

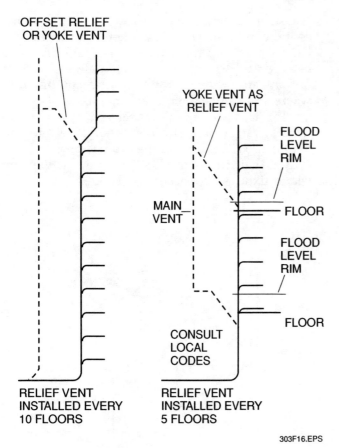

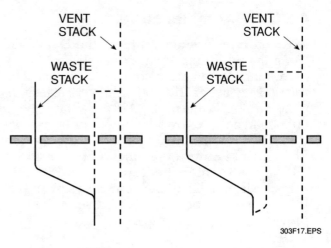

Figure 17 ◆ Two ways of constructing a relief vent using an offset.

Figure 16 ◆ Relief vent.

In multistory buildings, a relief vent can often be combined with an offset in the waste stack to slow down the waste (see *Figure 17*). The installation will depend on the design of the stack. Local codes will specify the appropriate use of an offset relief vent in your area.

4.7.0 Sovent® Vent Systems

An innovative vent design called the **Sovent® system** combines vent and waste stacks into one stack (see *Figure 18*). Where a branch joins the single stack, an aerator mixes waste from the branch with the air in the stack. This slows the velocity of the resulting mixture and keeps the stack from being plugged by wastewater. At the base of the stack, a de-aerator separates the air and the liquid. This relieves the pressure in the stack and allows the waste to drain out of the system. The Sovent® system is intended mostly for use in high-rise buildings.

 DID YOU KNOW?
Sovent® Vent Systems

Swiss professor Fritz Sommer invented a combination waste/vent system in 1959. Extensive tests proved the soundness of the basic design. Within a few years, Sovent® systems appeared in new buildings in Europe and the United States. Cast iron, hubless fittings appeared in 1977. These cost much less than previous copper pipe and fittings. Advocates of Sovent® systems praise their simplicity, efficiency, and quiet operation.

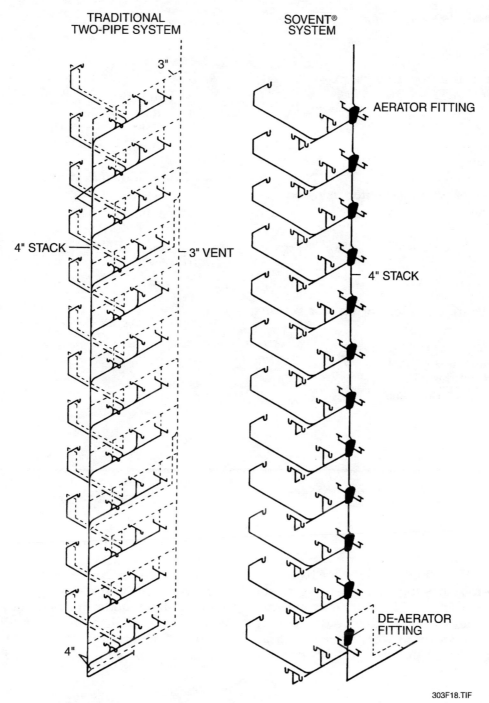

Figure 18 ◆ Two-pipe system compared to Sovent® system.

303F18.TIF

Review Questions

Section 4.0.0

1. _____ vents run from a single fixture directly to the vent stack.
   a. Branch
   b. Individual
   c. Continuous
   d. Common

2. Fixture batteries can be vented using either a circuit vent or a _____ vent.
   a. loop
   b. relief
   c. wet
   d. line

3. Wet vents are a popular option for venting _____.
   a. commercial water closets
   b. commercial lavatories
   c. island fixtures
   d. residential bathrooms

4. A yoke vent is a type of _____ vent.
   a. battery
   b. individual
   c. relief
   d. island

5. To slow down flow in a waste stack, combine a relief vent with a(n) _____.
   a. aerator
   b. offset
   c. branch interval
   d. re-vent

Summary

Vents allow air to enter and exit a DWV stack. Vents may ventilate single fixtures or a battery of them. They may provide relief venting to the main stack or connect with other vents. Regardless of how they are used, their purpose is to enable DWV systems to work by keeping an equal pressure inside the system. Vents come in many forms, and many can be used in combination with each other.

Like other plumbing installations, vents must be installed with proper grade and the correct pipe. Many model codes differ as to how vents should be installed and where different types of vents should be used. Always consult local codes and building plans prior to installing vents.

Trade Terms Introduced in This Module

Air admittance vent: A valve-type *vent* that ventilates a stack using air inside a building. The vent opens when exposed to reduced pressure in the vent system, and it closes when the pressure is equalized.

Back pressure: Excess air pressure in *vent* piping that can blow trap seals out through fixtures. Back pressure can be caused by the weight of the water column or by downdrafts through the *stack vent*.

Back vent: An *individual vent* that is installed directly at the back of a single fixture and connects the fixture drain pipe to the *vent stack* or the *stack vent*.

Branch interval: The space between two branches connecting with a main stack. The space between branch intervals is usually story height but cannot be less than 8 feet.

Branch vent: A *vent* that connects fixtures to the *vent stack*.

Circuit vent: A *vent* that connects a battery of fixtures to the *vent stack*.

Combination waste and vent system: A line that serves as a *vent* and also carries wastewater. *Sovent®* systems are an example of a combination waste and *vent* system.

Common vent: A *vent* that is shared by the traps of two similar fixtures installed back-to-back at the same height. It is also called a unit vent.

Continuous vent: A vertical continuation of a drain that ventilates a fixture.

Crown vent: A condition where a *vent* opening is located within two pipe diameters of a trap. Codes prohibit crown venting.

Flat vent: A *vent* that runs horizontally to a drain. It is also called a horizontal vent.

Hydraulic gradient: The maximum degree of allowable fall between a trap weir and the opening of a *vent* pipe. The total fall should not exceed one pipe diameter from weir to opening.

Indirect or momentum siphonage: The drawing of a water seal out of the trap and into the waste pipe by lower-than-normal pressure. It is the result of discharge into the drain by another fixture in the system.

Individual vent: A *vent* that connects a single fixture directly to the *vent stack*. *Back vents* and *continuous vents* are types of individual vents.

Loop vent: A *vent* that connects a battery of fixtures with the *stack vent* at a point above the waste stack.

Relief vent: A *vent* that increases circulation between the drain and *vent stacks*, thereby preventing *back pressure* by allowing excess air pressure to escape.

Re-vent: To install an *individual vent* in a fixture group.

Self-siphonage: A condition whereby lower than normal pressure in a drain pipe draws the water seal out of a fixture trap and into the drain.

Sovent® system: A *combination waste and vent system* used in high-rise buildings that eliminates the need for a separate *vent stack*. The system uses aerators on each floor to mix waste with air in the stack and a de-aerator at the base of the stack to separate the mixture.

Stack vent: An extension of the main soil and waste stack above the highest horizontal drain. It allows air to enter and exit the plumbing system through the building roof and is a terminal for other *vent* pipes.

Vent: A pipe in a DWV system that allows air to circulate in the drainage system, thereby maintaining equalized pressure throughout.

Vent stack: A stack that serves as a building's central *vent*, to which other vents may be connected. The vent stack may connect to the *stack vent* or have its own roof opening. It is also called the main vent.

Vent terminal: The point at which a *vent* terminates. For a fixture vent, it is the point where it connects to the *vent stack* or *stack vent*. For a *stack vent*, it is the point where the vent pipe ends above the roof.

Wet vent: A *vent* pipe that also carries liquid waste from fixtures with low flow rates. It is most frequently used in residential bathrooms.

Yoke vent: A *relief vent* that connects the soil stack with the *vent stack.*

Additional Resources

This module is intended to present thorough resources for task training. The following reference works are suggested for further study. These are optional materials for continued education rather than for task training.

Estimator's Man-Hour Manual on Heating, Air Conditioning, Ventilating, and Plumbing, 1978. John S. Page. Woburn, MA: Gulf Professional Publishing Company.

Planning Drain, Waste & Vent Systems, 1990. Howard C. Massey. Carlsbad, CA: Craftsman Book Company.

Plumbers and Pipefitters Handbook, 1996. William J. Hornung. Englewood Cliffs, NJ: Prentice Hall College Division.

References

International Association of Plumbers and Mechanical Officials (IAPMO) Web site, www.iapmo.org, "2000 International Plumbing Code/Uniform Plumbing Code Review," Edward Saltzberg and J. Richard Wagner, www.iapmo.org/common/pdf/upc_ipc.pdf, viewed August 2001.

Figure Credits

Copper Development Association, Inc.	303F17
Josam Manufacturing Company	303F07
Studor, Inc.	303F13

NCCER CRAFT TRAINING USER UPDATES

The NCCER makes every effort to keep these textbooks up-to-date and free of technical errors. We appreciate your help in this process. If you have an idea for improving this textbook, or if you find an error, a typographical mistake, or an inaccuracy in the NCCER's Craft Training textbooks, please write us, using this form or a photocopy. Be sure to include the exact module number, page number, a detailed description, and the correction, if applicable. Your input will be brought to the attention of the Technical Review Committee. Thank you for your assistance.

Instructors – If you found that additional materials were necessary in order to teach this module effectively, please let us know so that we may include them in the Equipment and Materials list in the Instructor's Guide.

Write: Curriculum Revision and Development Department
National Center for Construction Education and Research
P.O. Box 141104, Gainesville, FL 32614-1104

Fax: 352-334-0932

E-mail: curriculum@nccer.org

Craft _____   Module Name _____

Copyright Date _____   Module Number _____   Page Number(s) _____

Description _____

_____

_____

_____

(Optional) Correction _____

_____

_____

(Optional) Your Name and Address _____

_____

_____

Indirect and Special Waste

COURSE MAP

This course map shows all of the modules in the third level of the Plumbing curriculum. The suggested training order begins at the bottom and proceeds up. Skill levels increase as you advance on the course map. The local Training Program Sponsor may adjust the training order.

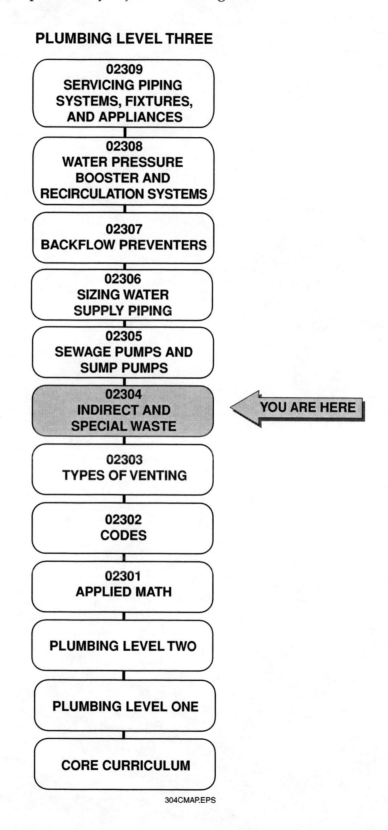

PLUMBING LEVEL THREE

02309
SERVICING PIPING
SYSTEMS, FIXTURES,
AND APPLIANCES

02308
WATER PRESSURE
BOOSTER AND
RECIRCULATION SYSTEMS

02307
BACKFLOW PREVENTERS

02306
SIZING WATER
SUPPLY PIPING

02305
SEWAGE PUMPS AND
SUMP PUMPS

02304
INDIRECT AND
SPECIAL WASTE ◄ YOU ARE HERE

02303
TYPES OF VENTING

02302
CODES

02301
APPLIED MATH

PLUMBING LEVEL TWO

PLUMBING LEVEL ONE

CORE CURRICULUM

304CMAP.EPS

MODULE 02304 CONTENTS

Figures

Table

Indirect and Special Waste

Objectives

When you have completed this module, you will be able to do the following tasks in accordance with local codes:

1. Identify the reasons for using indirect systems.
2. Discuss the requirements for receptors and backflow preventers.
3. Demonstrate the ability to install an indirect waste system.
4. Identify the reasons for using special waste systems.
5. Describe the purpose of interceptors and how each type functions.
6. Sketch the basic installation and maintenance requirements for interceptors.
7. Describe the precautions that must be taken when installing interceptors to ensure ease of future maintenance and repair.
8. Install an interceptor.
9. Use the local plumbing code to cite the requirements for using indirect waste disposal systems.
10. Use the local plumbing code to cite the requirements for using special waste disposal systems.

Prerequisites

Before you begin this module, it is recommended that you successfully complete the following: Core Curriculum; Plumbing Level One; Plumbing Level Two; Plumbing Level Three, Modules 02301 through 02303.

Required Trainee Materials

1. Appropriate personal protective equipment
2. Pencil and paper
3. Copy of your local code

1.0.0 ◆ INTRODUCTION

Reliable waste drainage is essential for many activities that we take for granted. When plumbing fails, restaurants are forced to close for health code violations, diseases spread in hospitals and hotels, and people can become ill and even die from exposure to toxic substances.

Plumbers design drain, waste, and vent (DWV) systems with two important considerations in mind:

- The systems must prevent sewer gas and waste from contaminating fixtures and appliances.
- The systems must prevent certain types of harmful wastes from entering the sewer system.

Fixtures that are designed to receive wastes, such as toilets, can handle contamination from sewage backups. Fixtures that are designed to hold food products, such as icemakers and food preparation sinks, must stay free of contamination. Otherwise, diseases may be passed on to the public.

2.0.0 ◆ INDIRECT WASTE SYSTEMS

All fixtures and appliances drain away waste products and thus become unsanitary. Fixtures are designed to be cleaned. Certain public use fixtures need extra protection from sewage backup. If a backup occurred in a restaurant ice bin, for example, it could cause illness and even death. One way to protect fixtures and appliances is to have them drain through pipes that are not connected to the main waste stack. Such pipes are called **indirect waste systems**. Indirect waste systems use **backflow preventers** to keep sewer wastes from contaminating the fixture if the drain backs up. An **air gap** is a commonly used backflow preventer. It is simply a space between the indirect waste disposal pipe and the main waste system (see *Figure 1*). The gap must be two times the diameter of the indirect waste disposal pipe, and it must not go into the air line.

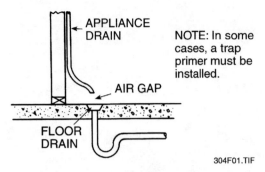

Figure 1 ◆ Indirect waste pipe with air gap.

The **air break** is another type of backflow preventer. In an air break, a smaller diameter pipe drains into a larger diameter pipe. The air break prevents the water in the larger pipe from returning to the smaller pipe. Use air breaks where there is no possibility of siphonage, such as in a lavatory. Washing machines are often installed with air breaks. Codes vary on the requirement for air gaps and air breaks. Always consult your local code before installing either one.

The wastewater that flows through indirect waste pipes is called **indirect waste**. Indirect wastes are relatively free of solids. Indirect wastes are produced by appliances and fixtures in:

- Commercial food preparation facilities
- Food storage areas
- Medical and dental offices
- Air conditioning condensate and other potable and nonpotable clear water waste installations

Indirect waste systems have many features in common. They also have their own special requirements.

2.1.0 Installation of Indirect Waste Piping

The air gap between an appliance waste pipe and the main drain must be at least twice the diameter of the appliance waste pipe. In other words, a waste pipe of two inches in diameter must have an air gap of at least four inches. Some indirect waste pipes empty into a sink basin. Measure the air gap from the flood level of the sink to the terminal of the waste pipe.

To prevent splashing, install a receptor on the inlet to the main waste system. Receptors can be funnels, floor drains, trapped and vented slop sinks, or other similar fixtures. Ensure that the receptor has a removable strainer or basket that will keep solids out of the waste pipe. Do not install receptors in bathrooms or in unventilated spaces, such as storerooms. Make sure that receptors are easily accessible for cleaning. An **open hub waste receptor** is a pipe or pipe hub extending at least one inch above the floor. This receptor does not require a strainer.

Another way to prevent splashing is to install a **modified waste line**. Using a sanitary tee, attach a vertical pipe to the terminal of the waste line (see *Figure 2*). The pipe should be open at both ends. A modified waste line allows the indirect waste to mix with air as it exits the pipe. This helps the wastewater to flow smoothly into the drain.

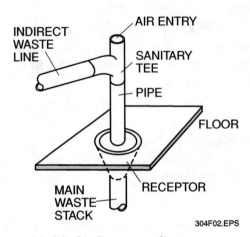

Figure 2 ◆ Modified indirect waste line.

Many indirect waste systems are used only intermittently. They may drain only small amounts of waste at a time. As a result, trap seals can dry up if not periodically refreshed. If that happens, sewer gases can escape into the surrounding air, causing odors and presenting a potential health hazard. Use an **automatic trap primer** to maintain a constant seal in a low-use trap (see *Figure 3*). To prevent the buildup of odors from sewer gases, ensure that indirect waste installations can be easily flushed. Install a hot water connection on longer lengths of waste pipe. A hose can be used to flush pipes not equipped with a hot water connection.

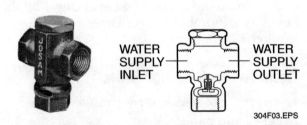

Figure 3 ◆ Automatic trap primer.

Codes prohibit the connection of steam lines to a plumbing system. Discharge the following wastes into a sump or basin connected to the cold water supply:

- Water that is hotter than 140°F (60°C)
- Condensates
- Hot water wastes
- Other hot liquids

Cold water lowers the wastewater temperature. After the wastewater has cooled, it is discharged into the sewer system.

2.1.1 Food Preparation, Handling, and Storage Installations

Codes require indirect waste lines with air gaps for walk-in refrigerators and freezers, ice makers, and ice compartments. Wastewater from multiple ice compartments and icemakers can be collected at one point (see *Figure 4*). This reduces the amount of piping required to drain multiple fixtures. Many codes require indirect waste lines that are longer than two feet to have traps.

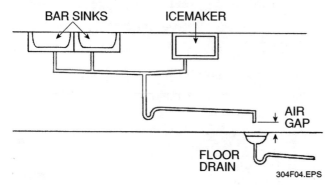

Figure 4 ◆ Indirect waste piping for bar sinks and icemakers.

Some codes allow commercial dishwashers and dishwashing sinks to attach directly to the main drain. If your code does not permit this, use an indirect waste system. Because large quantities of grease may be discharged from dishwashing fixtures, install a grease interceptor on the indirect waste line before the air gap. Grease and grease interceptors are discussed in more detail later in this module. Refer to your local plumbing code for the requirements in your area.

Many commercial kitchen appliances, such as potato peelers, create solid wastes. Install floor drains with **sediment buckets** to trap solid wastes (see *Figure 5*). A sediment bucket is a removable basin inside a drain that holds solid waste but allows water to drain.

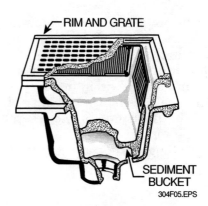

Figure 5 ◆ Floor drain with sediment bucket.

2.1.2 Health Care Installations

Health care facilities need sanitary conditions to protect the health of their patients. Effective waste disposal helps to limit the spread of disease. Indirect waste systems are used in a wide variety of health care facilities, including:

- Surgical and dental theaters
- Clinics and infirmaries
- Pharmaceutical and research laboratories
- Youth and elderly care institutions

Consult your local code before installing health care waste systems.

Install traps on indirect waste lines for bedpan steamers. Batteries of fixtures, such as sterilizers, can drain into a single **receptor**. A receptor is a basin that catches and holds liquids. Attach clinical sinks directly to the main waste system. Bedpan washers can also be connected directly to the waste system. They should be vented.

 WARNING!
Wastewater from medical and dental facilities is hazardous. It can contain bacteria and viruses that can cause illness and death. Ensure that the design of the DWV system adheres strictly to code before installing it.

2.1.3 Other Clear Water Waste Installations

Many other fixtures and appliances create indirect wastes. Installations that produce clear water wastes include:

- Swimming pools
- Relief valves
- Air conditioner condensate drains
- Refrigerators
- Water storage tanks
- Water heaters
- Drinking fountains

Water heater relief lines create indirect waste only on occasion. Install an automatic primer to maintain a constant trap seal (see *Figure 6*). Drinking fountains often drain into a single receptor and should drain into a receptor with a strainer (see *Figure 7*). Measure the air gap between the flood line of the receptor and the waste pipe terminal. Refer to your local code when installing these and other indirect waste-water systems.

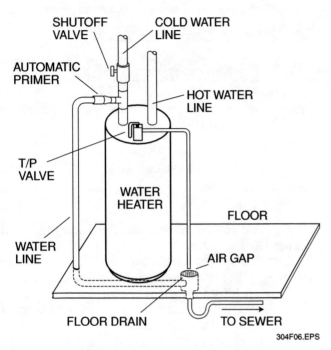

Figure 6 ◆ Indirect waste line on a water heater.

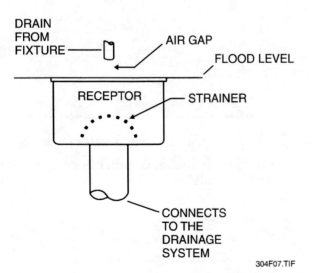

Figure 7 ◆ Indirect waste line for a drinking fountain.

Review Questions

Section 2.0.0

1. Air gaps and air breaks are commonly used _____.
    a. indirect waste lines
    b. floor drains
    c. backflow preventers
    d. waste receptors

2. The air gap between an indirect waste drain and the main waste pipe must be _____ the drainpipe diameter.
    a. at least two times
    b. one-half
    c. at least four times
    d. equal to

3. Indirect wastes can contain _____ solid matter.
    a. small amounts of
    b. large amounts of
    c. no amount of
    d. a fixed percentage of

4. To prevent buildup of odors from sewer gas, install a _____ on longer lengths of waste pipe.
    a. sediment bucket
    b. sanitary tee
    c. fixture trap
    d. hot water connection

5. A receptor is a basin that catches and _____ liquids.
    a. discharges
    b. holds
    c. drains
    d. cleans

3.0.0 ◆ SPECIAL WASTE SYSTEMS

Indirect waste is not the only kind of waste that requires special drain installations. **Special wastes** must be treated before they can discharge safely into the system for a number of reasons:

*   Their temperature is too high for the septic or sewer system, such as water above 140°F.
*   They contain solid and partially solid matter that will settle out.
*   They contain acids that must be neutralized.
*   They contain harmful or potentially harmful substances, such as grease, oil, and gasoline.

The Trap Trap

Interceptors are sometimes called *traps*, such as a grease trap. You have already learned that the term trap also applies to installations that maintain water seals in fixtures. Both types of traps are vital components in DWV systems because they protect people from exposure to noxious and harmful substances. However, they operate very differently. Interceptors and traps have their own unique requirements. Do not confuse the two! The results could be dangerous. Take the time to review your local code requirements for both interceptor and trap installations.

Some special wastes are treated by evaporation, while others must be separated from the wastewater. Still others have to be diluted or chemically neutralized first.

3.1.0 Types of Special Waste

Special wastes include grease, oil, and gasoline; solids, such as sediment, glass, and hair; corrosive wastes; and high-temperature wastes. Grease or animal fat is generated in large quantities where food is prepared and served commercially. Grease will clog sewer and septic pipes if it is allowed to enter the system. Oil and gasoline are volatile petroleum products. They are used to operate, clean, repair, and store motor vehicles. These fluids could ignite and damage sewer, septic, or storm water drainage systems. Sediment can clog drains and pipe. Corrosive chemicals like acids can damage pipes and drains, generate toxic fumes, and interfere with the waste treatment process. High-temperature wastes can cause pipes to expand and contract. This weakens pipe joints and causes leaks.

Treatments vary for each type of waste. Improper treatment could damage the sewer or septic system. It could also contaminate fresh water supplies and cause sickness or even death. Review the proper treatment method before treating wastes. Your local code will also provide specific guidelines.

3.2.0 Installation of Interceptors

Interceptors trap special wastes before they can enter the sewer system. Small interceptors are about the size of an ordinary P-trap. The largest custom-built underground interceptors are large enough for a person to stand in. No matter what size they are, most interceptors operate the same basic way.

The internal volume of an interceptor is larger than that of the drainage pipes. This causes wastewater to slow down as it enters, allowing the waste materials time to settle, float, or evaporate out of the wastewater. **Baffles** are partitions in the interceptor that reduce turbulence caused by flowing wastewater. Baffles also keep wastes from escaping through the outlet. Free of the contaminants, the wastewater discharges into the sewer system by gravity.

Interceptors work a lot like fixture traps. Both are designed to seal one part of a plumbing installation from waste products in another part. However, interceptors violate one of the basic principles of traps: interceptors are not self-cleaning and must be manually cleaned.

Site plans specify where large interceptors should be installed. An interceptor can be above or below floor level. The installation's design and the available space determine the location. Provide enough space to allow cleaning equipment, such as pumps and augers, to fit. Ensure that covers and sediment buckets can be removed easily.

Locate the interceptor close to the fixture because this will help prevent the inlet piping from clogging. Install exterior interceptors below the frost line to keep them from freezing. Support large interceptors with a solid foundation because the weight of aboveground interceptors can cause structural problems in some buildings. Even interceptors installed below grade need support. Otherwise, pipe joints may fail and damage the tank. Packed sand over undisturbed earth is a good foundation.

Maintaining Special Waste Interceptors

Special waste interceptors require regular service. Provide the building managers or occupants with a maintenance schedule. Explain that regular servicing protects the health of the public and helps protect the owners from fines, penalties, and lawsuits.

If the waste includes toxic gases, steam, high-pressure wastes, or chemicals, vent the interceptor through a roof vent. Ensure that the inlet and outlet piping is installed properly, and seal all piping joints tightly to prevent liquid or gas leaks. For concrete and masonry interceptors, carefully grout the joints between the piping and the tank.

3.2.1 Grease Interceptors

For small amounts of grease, install a small interceptor inside the building (see *Figure 8*). They are typically made of steel and have a bolt-on lid. Ensure that the interceptor is easy to reach for cleaning, and locate it as close to the sink or appliance as possible. Otherwise foul odors may occur. Odors are the biggest problem associated with indoor grease interceptors. Proper venting will help reduce odors. Vent grease interceptors using the guidelines for venting a trap. Cleaning can be a problem because the lid must be removed.

One cleaning option is to use a **draw-off hose** (see *Figure 9*). Draw-off hoses are used to siphon the grease out of an interior grease interceptor; they can be either gravity fed or connected to a pump. Ensure that a draw-off valve is installed at the outlet of the interceptor. Closing the valve allows the water level in the interceptor to rise, which forces the grease to flow into the draw-off hose. Consult your local code for the proper procedure.

Excessive flow into an interceptor can cause turbulence and reduces the amount of time grease has to coagulate and float to the top for draw-off. These conditions will permit grease to escape through the outlet. Install a **flow control fitting** on the inlet pipe to slow the flow of wastewater into an interceptor (see *Figure 10*). The flow control's inlet is carefully sized, limiting the amount of wastewater that can flow into the interceptor. Flow control devices with removable or adjustable parts are not permitted in grease interceptors.

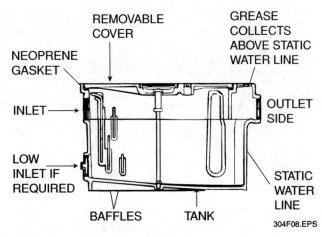

Figure 8 ◆ Typical interior grease interceptor.

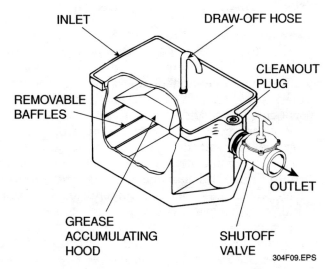

Figure 9 ◆ Draw-off hose installed on a grease interceptor.

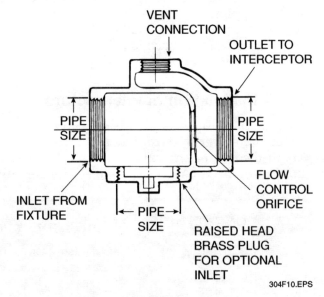

Figure 10 ◆ Flow control fitting.

DID YOU KNOW?

The First Restaurant

In 1765, a cook named Boulanger opened a store in Paris that sold a selection of soups and broths. He called his soups *restoratives* or, in French, *restaurants*. Other chefs soon opened similar stores throughout Europe. They also called their stores restaurants. The term came to denote an eating establishment where people can select from a menu.

Large interceptors are located outside and underground. Precast concrete tanks, similar to septic tanks, can be used as large grease interceptors. They can be site-built from concrete, masonry, and pipe fittings (see *Figure 11*). The dimensions can be varied to meet size requirements. Ensure that the inlet is positioned correctly with regard to the outlet and the static water line. Increasing the height of the outlet pipe allows more solids to collect. This reduces the number of times the interceptor has to be serviced. Access to the interceptor is provided by a manhole or through an opening for a draw-off hose. Two- and three-chamber grease interceptors are more efficient than single-chamber designs (see *Figure 12*).

Codes provide formulas to allow you to calculate the right size for grease traps. The formulas in *Table 1* are based on the Environmental Protection Agency (EPA)-2 Model; always refer to your local code. First, determine the total maximum drainage flow from each fixture. This number is estimated in gallons per minute. Next, factor in the estimated load factors. Multiply the total drainage flow by the load factor. Multiply that number by 60 (minutes) to get the maximum flow in one hour. Finally, multiply the maximum flow by 2 (representing the two-hour retention time specified in the code). This number gives you the total required volume of the grease trap in gallons.

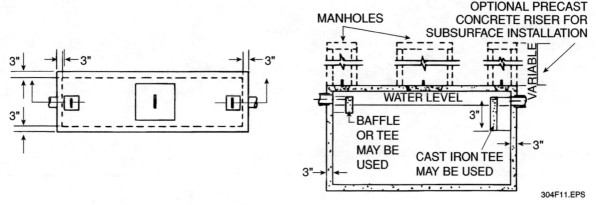

Figure 11 ◆ Typical large site-built grease interceptor.

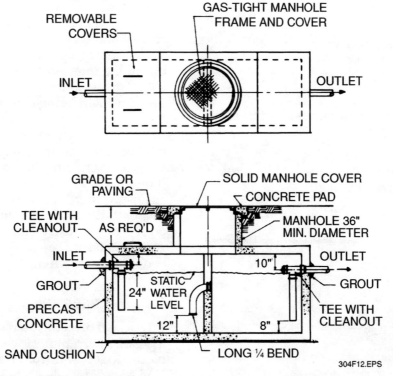

Figure 12 ◆ Two-chamber grease interceptor.

Table 1 Grease Interceptor Sizing Formula for a Restaurant

A. Determine maximum drainage flow from fixtures:		
Type of Fixture	**Flow Rate**	**Amount**
Restaurant kitchen sink	15 gpm	_____
Single compartment sink	20 gpm	_____
Double compartment sink	25 gpm	_____
2 single compartment sinks	25 gpm	_____
2 double compartment sinks	35 gpm	_____
Triple sink, 1½-inch drain	35 gpm	_____
Triple sink, 2-inch drain	35 gpm	_____
30-gallon dishwasher	15 gpm	_____
50-gallon dishwasher	25 gpm	_____
50- to 100-gallon dishwasher	40 gpm	_____
B. Total		_____
C. Estimate Loading Factors:		
Restaurant type Fast food/paper delivery	=	.50
Low volume	=	.50
Medium volume	=	.75
High volume	=	1.00
D. Step B × Step C: _____	(Subtotal)	
E. Subtotal × 60 min. = max. flow for 1 hour		_____
F. Step E × 2 hours' retention time = vol. of trap (gal.)		_____

To clean small grease interceptors, use the following procedure:

Step 1   Remove the lid.

Step 2   Scoop the grease out of the interceptor and place into a watertight container (wear proper protective equipment for the hands and face).

Step 3   Thoroughly clean the sides and inside lid of the interceptor.

Step 4   Make sure the container is tightly sealed before properly disposing of it according to local code requirements.

Step 5   Re-install and seal the lid.

WARNING!

Climb-in interceptors may contain corrosive, flammable, or odorous substances. Personal injury could result in poisoning or an infection, such as hepatitis. Obtain thorough confined-space training before servicing climb-in interceptors. Use appropriate personal protective equipment, and wear hand protection while working.

Grease disposal varies from city to city. Some areas forbid placing grease in landfills. Other codes allow some grease to be placed in garbage cans in sealed containers. Professional contractors perform service and repair on exterior grease interceptors. They use large pumper trucks to suck the grease out of the interceptor. Some recycling companies collect grease and convert it into fuel and industrial products. Rendering companies also collect grease and use it in the manufacture of consumer products, such as soap and perfume. As a plumber, you may be called on to recommend contractors, recyclers, and rendering companies. Only recommend those that follow environmental standards.

Ensure that food-waste grinders and garbage disposals do not discharge into a grease trap. Most codes place restrictions on garbage disposal wastes entering sewer systems. Check your local code for the requirements.

3.2.2 Oil Interceptors

The design of an oil interceptor is similar to that of a grease interceptor. However, oil and gasoline are harder to separate from wastewater than grease is. Codes vary widely on how to install oil interceptors. You may need to consult a local code authority.

Use draw-off hoses to remove oil from an interceptor. Otherwise, oil will have to be skimmed out by hand—a messy operation. There are two ways to use draw-off hoses. **Gravity draw-off systems** (*Figure 13*) drain oil into a storage tank. The oil drains out when it reaches a certain height within the interceptor. The height is determined by an adjustable draw-off sleeve inside the interceptor. For gravity systems, set the terminal of the sleeve to ⅛ inch above the normal water operating line. In **manual draw-off systems** (*Figure 13*), oil drains out of the interceptor by operating a draw-off valve. The terminal of the adjustable draw-off sleeve must be set to ¼ inch below the static water line. Be sure to close the flow control valve before operating the draw-off valve. Both draw-off systems use sediment buckets or similar devices to catch dirt and grit before they enter the drain lines.

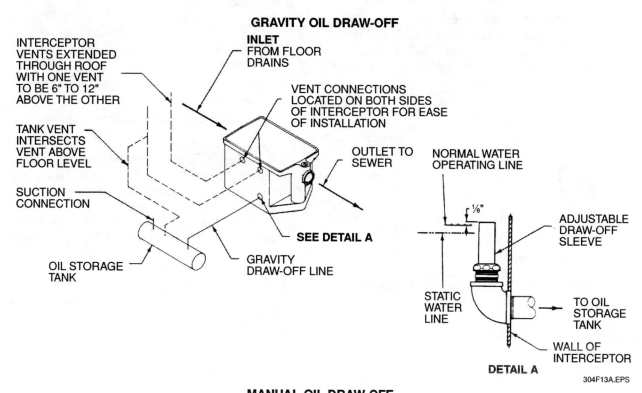

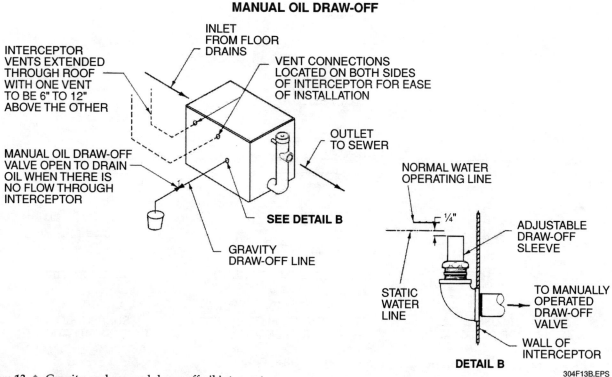

Figure 13 ◆ Gravity and manual draw-off oil interceptors.

Install vents on oil interceptors. Vents allow flammable and noxious gases to escape harmlessly and prevent back pressure from building up in the interceptor. Oil interceptors require two vents; terminate one vent 6 to 12 inches above the other. Oil storage tanks on gravity draw-off systems must be vented.

3.2.3 Sediment Interceptors

Sediment interceptors prevent sand, dirt, and grit from entering the waste system (see *Figure 14*). Small interceptors are installed in place of the P-trap. The top and bottom of sediment interceptors are designed to be removed. This allows for regular cleaning no matter how it is installed. Large sediment interceptors can be installed below, on, or above the floor (see *Figure 15*). They can be either precast or site-built (see *Figure 16*). Larger inlets may require the installation of a vent.

Specially designed sediment interceptors can also be installed in a variety of other facilities, including:

- Car washes
- Parking garages
- Swimming pools
- Commercial laundries (*Figure 17*)
- Jewelry shops
- Computer chip manufacturing plants

Each type of interceptor has its own special requirements. Be sure to talk with an experienced plumber before you put in a special interceptor. Review your local code as well.

Provide enough space for a drain pan to fit below the interceptor. This will prevent liquid wastes from spilling when the interceptor is cleaned. On small sediment interceptors, the lower opening should be the inlet. Otherwise, the interceptor will not hold enough liquid to form a trap seal. In that case, sewer gas can escape and special wastes can enter the sewer or septic system.

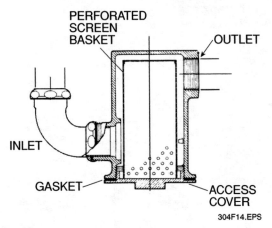

Figure 14 ◆ Small sediment interceptor.

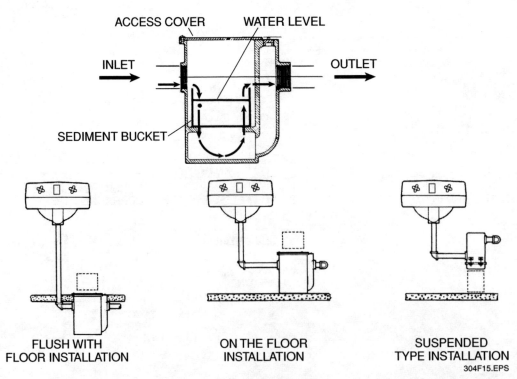

Figure 15 ◆ Large sediment interceptors installed on wall-hung hand sinks.

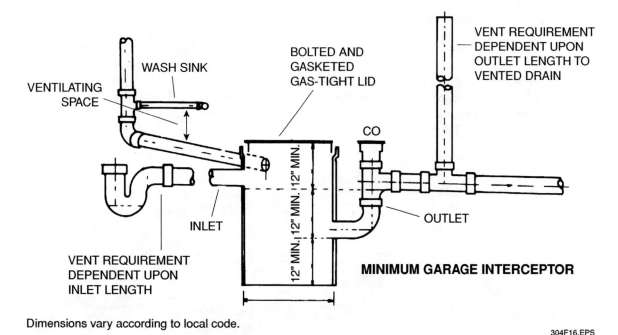

Dimensions vary according to local code.

304F16.EPS

Figure 16 ◆ Example of a site-built sediment interceptor with typical information provided in design drawings.

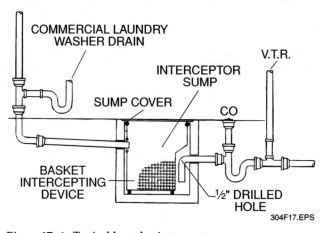

304F17.EPS

Figure 17 ◆ Typical laundry interceptor.

3.2.4 Catch Basins

As you have seen, sediment buckets installed in floor drains intercept sand and grit deposited on the floor. **Catch basins** perform a similar function, but on a larger scale. Catch basins are reservoirs that allow sediment to settle before wastewater discharges into the drain system. Install catch

CAUTION

Do not use catch basins or sediment interceptors to trap sanitary wastes. Install traps on all interceptors that discharge into sanitary sewer systems. This will allow the interceptor to drain properly.

basins in parking lots, roof drains, and yard drains. Catch basins can be customized to match installation needs. Review local codes for applicable requirements.

Catch basins are also used to cool high-temperature wastes before they enter the waste system. Do not install catch basins to intercept sanitary sewage. Catch basins do not require traps if they drain into a storm sewer or drainage field. Ensure that cover plates on the catch basin are easy to access so that built-up sediment can be removed.

3.3.0 Installation of Neutralization Sumps

Corrosive liquids, spent acids, and toxic chemicals pose serious health risks. They can't discharge directly into the drainage system. They must first be diluted, chemically treated, or neutralized. **Neutralization sumps**, or tanks, are used to treat these types of wastes. Only then can these wastes discharge into the sewer. Consult your local code for specific requirements.

Neutralize the intermittent flows of some wastes by diluting them. Ensure that the sump is large enough to permit the waste to dilute to an acceptable level according to local code. Refer to your local code for chemicals that can be treated through dilution. Sumps used for dilution must be vented. Venting allows toxic gas to escape safely. Chemical waste and sanitary vents must extend through the roof.

Some wastes must pass through a **neutralizing agent** before they can be discharged (see *Figure 18*). Neutralizing agents react chemically with special wastes. The reaction turns the wastes into substances that can be discharged safely. One of the most commonly used agents is limestone. Sumps with neutralizing agents must be vented because venting allows the gases resulting from the chemical reaction to escape. Install gas-tight lids on sumps with neutralizing agents, and ensure that the sump contains enough of the neutralizing agent to treat all the special waste that will flow through it. Sumps must be accessible so that the neutralizing agent can be replaced.

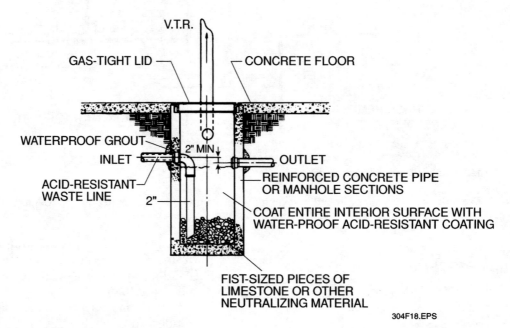

304F18.EPS

Figure 18 ◆ Neutralizing sump with neutralizing agent.

Review Questions

Section 3.0.0

1. Upon entering an interceptor, wastewater _____.
   - a. accelerates
   - b. decelerates
   - c. heats
   - d. evaporates

2. Interceptors violate one of the basic principles of traps because they _____.
   - a. are not self-cleaning
   - b. do not have moving parts
   - c. come into contact with toxic substances
   - d. are available in different sizes

3. Packed sand over _____ provides a solid foundation for large exterior interceptors.
   - a. a concrete slab
   - b. packed earth
   - c. undisturbed earth
   - d. packed gravel

4. Closing a grease draw-off valve to drain the grease from the interceptor _____ the water in an interceptor.
   - a. redirects the flow of
   - b. raises the level of
   - c. drains
   - d. lowers the level of

5. To remove oil from an interceptor using the gravity draw-off method, the terminal of the adjustable draw-off sleeve must be set to _____ the normal water operating line.
   - a. ¼ inch above
   - b. ¼ inch below
   - c. ⅛ inch below
   - d. ⅛ inch above

Summary

Indirect waste systems protect food preparation, serving and storage areas, health care facilities, and residential appliances from contamination by sewer wastes. Indirect waste connections with air gaps are installed on sinks, refrigerators and freezers, food preparation surfaces, and medical and dental equipment. Backflow preventers also help protect fresh water supplies from exposure to indirect wastes through cross-connection.

Special wastes must be treated before they can enter the waste system. Interceptors prevent grease, oil, and sediment from entering the sewer system. Chemicals are diluted and neutralized in sumps. High-temperature wastes cool in catch basins, which can also be used to trap sediment.

Specific installation requirements vary for indirect and special waste systems. Before installing such a system, consult your local code and take the time to discuss installation designs with code experts. Proper treatment of indirect and special wastes helps ensure the health and safety of the public.

Trade Terms Introduced
in This Module

Air break: A *backflow preventer* in which a smaller pipe drains into a larger pipe.

Air gap: A simple *backflow preventer* that consists of a space between the *indirect waste system* and the main waste system. The gap must be two times the diameter of the indirect waste disposal pipe and must not go into the air line.

Automatic trap primer: A device that feeds water into a low-use trap. It allows the trap to maintain a constant seal.

Backflow preventer: A device that protects fixtures and appliances from contaminated wastes. It is installed on a fresh-water line or a drain line.

Baffle: Partition inside the body of an *interceptor*. Baffles are designed to reduce turbulence caused by incoming wastewater and to block wastes from escaping through the outlet.

Catch basin: A reservoir installed before the sanitary piping. It allows sediment to settle out of wastewater.

Draw-off hose: A hose used to siphon grease and oil from *interceptors*.

Flow control fitting: A device that regulates the flow of wastewater into grease and oil *interceptors*.

Gravity draw-off system: A system that automatically drains oil from an *interceptor* into a storage tank.

Indirect waste: Wastewater that flows through an indirect waste pipe.

Indirect waste system: A drainpipe attached to a fixture or appliance that is separated from the regular drainage system by a *backflow preventer*.

Interceptor: A device that prevents *special waste* from entering the sewer or septic system.

Manual draw-off system: A system that drains oil out of an *interceptor* through a valve-operated *draw-off hose*.

Modified waste line: A vertical pipe at the terminal of a waste line that allows *indirect waste* to mix with air.

Neutralization sump: A tank where *special wastes* are mixed with a *neutralizing agent*.

Neutralizing agent: A substance, such as limestone, that reacts chemically with *special waste*. Neutralizing agents make special waste safe to empty into a sanitary system.

Open hub waste receptor: A waste drainpipe or pipe hub that extends above the floor line.

Receptor: A basin designed to hold liquids.

Sediment bucket: A removable basin in a drain that prevents solid wastes from entering a sanitary system.

Special waste: Waste that must be removed, diluted, or neutralized before it can be discharged into a sanitary system.

Additional Resources

This module is intended to present thorough resources for task training. The following reference works are suggested for further study. These are optional materials for continued education rather than for task training.

Introduction to Wastewater Treatment Processes, 1997. Rubens Sette Ramalho. New York, NY: Academic Press.

Planning Drain, Waste & Vent Systems, 1993. Howard C. Massey. Carlsbad, CA: Craftsman Book Company.

Water, Sanitary, and Waste Services for Buildings, 1995. Alan F. E. Wise and J. A. Swaffield. Boston, MA: Addison-Wesley.

Acknowledgments

Ken Swain, Plumb Works, Atlanta, GA.

"Grease: Is Grease a Problem?" Informational brochure, available at www.ci.kennewick.wa.us/pw/grscomm.htm. Contact: pwinfo@ci.kennewick.wa.us

Mr. Lawrence Jacobs, Editor, Craftsman Book Company, 6058 Corte del Cedro, Carlsbad, CA 92009, (800) 829-8123 or (760) 438-7828 ext. 304

Figure Credits

Josam Manufacturing Company   304F03, 304F05

Wade Division/Tyler Pipe   304F08, 304F09, 304F10, 304F13, 304F14, 304F15

NCCER CRAFT TRAINING USER UPDATES

The NCCER makes every effort to keep these textbooks up-to-date and free of technical errors. We appreciate your help in this process. If you have an idea for improving this textbook, or if you find an error, a typographical mistake, or an inaccuracy in the NCCER's Craft Training textbooks, please write us, using this form or a photocopy. Be sure to include the exact module number, page number, a detailed description, and the correction, if applicable. Your input will be brought to the attention of the Technical Review Committee. Thank you for your assistance.

Instructors – If you found that additional materials were necessary in order to teach this module effectively, please let us know so that we may include them in the Equipment and Materials list in the Instructor's Guide.

Write: Curriculum Revision and Development Department
National Center for Construction Education and Research
P.O. Box 141104, Gainesville, FL 32614-1104

Fax: 352-334-0932

E-mail: curriculum@nccer.org

Craft _____   Module Name _____

Copyright Date _____   Module Number _____   Page Number(s) _____

Description _____

_____

_____

_____

(Optional) Correction _____

_____

_____

(Optional) Your Name and Address _____

_____

_____

_____

Sewage Pumps
and Sump Pumps

COURSE MAP

This course map shows all of the modules in the third level of the Plumbing curriculum. The suggested training order begins at the bottom and proceeds up. Skill levels increase as you advance on the course map. The local Training Program Sponsor may adjust the training order.

PLUMBING LEVEL THREE

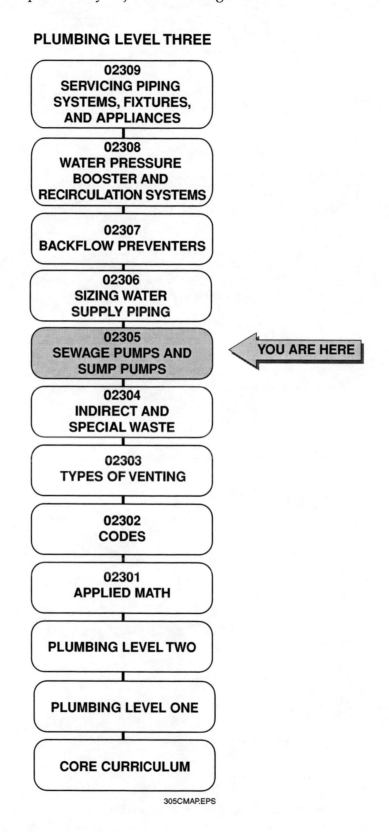

02309
SERVICING PIPING
SYSTEMS, FIXTURES,
AND APPLIANCES

02308
WATER PRESSURE
BOOSTER AND
RECIRCULATION SYSTEMS

02307
BACKFLOW PREVENTERS

02306
SIZING WATER
SUPPLY PIPING

02305
SEWAGE PUMPS AND
SUMP PUMPS

YOU ARE HERE

02304
INDIRECT AND
SPECIAL WASTE

02303
TYPES OF VENTING

02302
CODES

02301
APPLIED MATH

PLUMBING LEVEL TWO

PLUMBING LEVEL ONE

CORE CURRICULUM

305CMAP.EPS

MODULE 02305 CONTENTS

Figures

Tables

Sewage Pumps and Sump Pumps

Objectives

When you have completed this module, you will be able to do the following tasks in accordance with local codes:

1. Explain the functions, components, and operation of sewage and sump pumps.
2. Size a storm water sump by calculating the runoff from paved and unpaved land surfaces.
3. Size a sewage sump by calculating the sewage flow from a structure.
4. Install and adjust sensors, switches, and alarms in sewage and sump pumps.
5. Troubleshoot and repair sewage and sump pumps.
6. Using a detailed drawing, identify system components.
7. Install a sump pump.

Prerequisites

Before you begin this module, it is recommended that you successfully complete the following: Core Curriculum; Plumbing Level One; Plumbing Level Two; Plumbing Level Three, Modules 02301 through 02304.

Required Trainee Materials

1. Appropriate personal protective equipment
2. Pencil and paper
3. Copy of your local code

1.0.0 ◆ INTRODUCTION

Plumbing systems channel wastewater away for disposal through drains. Drains prevent contamination by sewage wastes and flooding by runoff water. You learned how to size and install drains in *Plumbing Level Two*. Often, the design of a building makes it necessary for a plumber to locate a drain below the sewer line. These drains are called **sub-drains**. Such drains can't discharge into the sewer by gravity. Instead, sub-drains must empty into temporary holding pits. Pumps installed in the pits provide the lift to move wastewater up out of the pit and into higher-level drainage lines. Storage pits and pumps that handle sewage wastes are called **sewage removal systems**. Storage pits and pumps that handle clear water runoff are called **storm water removal systems.**

This module reviews the components that plumbers use to build sewage and storm water removal systems. It also discusses how to troubleshoot and repair the most common problems with pumps, controls, and storage pits. Remember always to review your local code before designing, installing, or repairing pumps and other components. Sewage and storm water removal systems help maintain public health and safety. Plumbers are responsible for their safe operation and maintenance.

2.0.0 ◆ SEWAGE REMOVAL SYSTEMS

Sewage removal systems are often called **lift stations.** A complete sewage removal system consists of several components: sewage pumps, sumps, and controls. A **sewage pump** creates a partial vacuum that draws the waste from the storage pit into the pump and forces it out with sufficient pressure to reach the sewer above. Sewage pumps are often called **ejectors**. A **sump** collects sewage from the sub-drain. Sumps are also called *wet wells* or *sump pits*. Each installation has controls that operate the pump. Controls measure the amount of wastewater in the sump, turn the pump on and off, and provide backup in case of pump or power failure. There are two distinct types of sewage removal systems: centrifugal and pneumatic. The names refer to the type of pump used to remove the waste. Each of these systems and their components are discussed in more detail below.

2.1.0 Sewage Pumps

Two types of pumps are used in sewage removal systems: **centrifugal pumps** and **pneumatic ejectors.** Both types of pump are designed for specific applications. Be sure to review the design of the building's waste system. This will help you select the most appropriate type of pump. Always follow the manufacturer's instructions when installing pumps and ejectors. Join pumps to the inlet and outlet piping with either a union or a flange. This permits the pump to be easily removed for maintenance, repair, or replacement. The following installation steps are common for most sewage and storm water pumps:

Step 1 Before installing the pump in the basin, ensure the pump is appropriate for the sump.

Step 2 Check that the voltage and phase from the electrical supply matches the power requirements shown on the pump's nameplate.

Step 3 Thread the sump's discharge pipe into the pump's discharge connection.

Step 4 Install a check valve horizontally if one is required. If this is not possible, install it at a 45-degree angle with the pivot at the top to prevent solids from clogging the valve.

Step 5 Drill a ³⁄₁₆-inch (or recommended size) hole in the discharge pipe about 2 inches above the pump discharge connection to prevent air from locking the pump.

Step 6 If required, install a gate valve in the system after the check valve to permit removal of the pump for servicing.

Step 7 Install a union above the high water line between the check valve and the pump to allow the pump to be removed without disturbing the piping.

2.1.1 Centrifugal Pumps

Centrifugal pumps are frequently used in both storm water and sewage removal systems. Centrifugal pumps consist of the pump, an electric motor to run the pump, a plate on which to mount the pump, and controls to turn the pump on and off at the proper time. Many pumps come with motors that are designed to be submersible, which means the entire pump assembly can be placed inside the sump (see *Figure 1*). Others have a motor that must be installed above the sump, with the pump body connected to the motor by a sealed shaft (see *Figure 2*). Centrifugal pumps are widely used in sewage removal systems because of their mechanical simplicity and low cost. Complete centrifugal pump systems are available as self-contained units with pump, controls, and basin. They may also be fabricated on site from individual components.

Inside the body of a centrifugal pump is a set of spinning vanes called an **impeller.** An impeller removes wastewater from a sump in two steps. First, it draws wastewater up from the sump by

DID YOU KNOW?

History of the Pump

One of the earliest known pumps was the *shaduf*, developed in Egypt around 1550 B.C.E. The shaduf was a lever that lifted water from a well. About 400 years later in Persia, people pumped water using a chain of pots tied to a long loop of rope. The loop was wrapped around a wheel. As the wheel turned, the pots scooped water from a well and brought it to the top. About 650 B.C.E., Romans in Egypt pumped water with a similar device called a *noria*. This was a wheel with pots attached to the rim. When the wheel was lowered partly into the water and spun, the buckets dipped into the water. In 230 B.C.E., the Greek inventor Archimedes designed a pump made out of a coil of pipe wrapped around a shaft. By dipping one end of the pipe into the water and turning the shaft, the water corkscrewed its way up the pipe coil and came out the top end. Water wheels and Archimedean screws are still widely used throughout the world.

The Byzantines used a piston and cylinder to lift water using the force of air pressure in the first century B.C.E. Air pressure is also used in the centrifugal pump, which was invented by Leonardo da Vinci in the 1500s. It employs a set of spinning blades to force water into and out of a chamber. An engineer named Denis Papin designed the first practical centrifugal pump in 1688. The centrifugal pump is widely used in modern sewage and storm water removal systems.

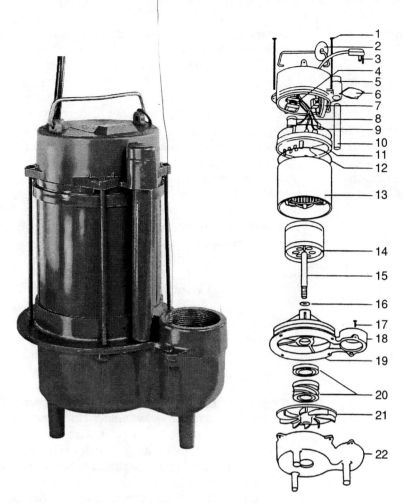

1	STAINLESS STEEL BOLTS
2	CARRYING HANDLE AND DATE TAG
3	POWER CORD
4	MOTOR START RELAY
5	MOTOR COVER
6	HUMISORB PACKET
7	WATER LEVEL SWITCH
8	OVERLOAD PROTECTOR
9	CONNECTING LEADS
10	PVC-AIR TUBE
11	BAFFLE AND TERMINAL PLATE
12	RUBBER GASKET
13	STATOR ASSEMBLY – 115V
14	ROTOR SHAFT ASSEMBLY
15	SHAFT
16	THRUST RACE
17	SCREW
18	CAST IRON MOTOR END
19	DIVERTER FLANGE
20	CERAMIC SEAT AND ROTARY SEAL HEAD
21	IMPELLER
22	HOUSING PLATE

305F01.EPS

Figure 1 ◆ Submersible centrifugal sewage pump.

creating a partial vacuum. It then ejects the swirling water out of the pump body into the discharge line. Centrifugal pumps are water-lubricated, which means that they must always be in contact with water or they will suffer damage. If this happens, repair or replace the pump. Ensure that centrifugal pumps are primed before installing them.

Centrifugal pumps can be designed with non-clogging and self-grinding features that efficiently handle solid wastes up to a certain size. The International Plumbing Code® (IPC), for example, requires that pumps receiving wastes from water closets must be able to handle solids up to 2 inches in diameter. Other pumps must be able to pump solids up to 1 inch in diameter. Because sand and grit in the wastewater can cause significant wear on pump components, install sediment interceptors in the drain line. Centrifugal pumps may require maintenance to replace moving parts like the impeller. Install pumps so that they can be removed easily from sumps.

Sometimes a design may call for a backup capability in case a pump fails, or a design may require extra pumping capacity during periods of peak flow. In such cases, install a **duplex pump** to fulfill

Figure 2 ◆ Centrifugal pump installation with aboveground motor.

305F02.TIF

the requirement. A duplex pump is simply two complete centrifugal pump assemblies, including floor plates, controls, and high water alarms, installed in parallel in a sump (see *Figure 3*). Duplex pumps can also be designed so that they operate on an alternating basis, reducing the wear on a single pump. Use an electrical pump alternator to switch between pumps in a duplex arrangement. Remember to size the sump to hold both pumps. Note also that sump covers must be specially ordered to accommodate a duplex installation.

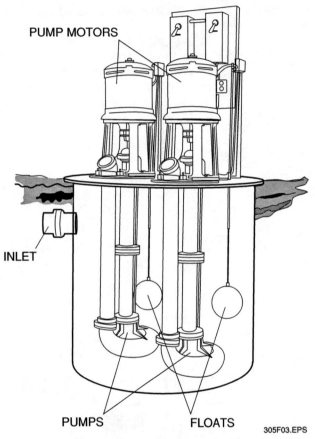

PUMP MOTORS

INLET

PUMPS   **FLOATS**   305F03.EPS

Figure 3 ◆ Duplex pump installation.

Rags and stringy solids may clog centrifugal pumps. Floating wastes and scum are out of the reach of a pump inlet that is located at the bottom of a sump. For these types of sewage wastes, use a **reverse flow pump** arrangement. Reverse flow pumps consist of a duplex pump installed with strainers, cutoff valves, and check valves (see *Figure 4*). Reverse flow pumps are designed to discharge solid wastes without passing them through the impeller housing. Because reverse flow pumps consist of two separate pumps, the system offers a safety backup in case one pump fails.

As wastewater enters a sump through one inlet, a strainer traps solids. When the water reaches a predetermined height, the pump switches on. The pump forces the trapped solids out of the strainer and through the discharge pipe. To prevent back pressure into the soil line when the pump is operating, install a full-opening, positive-seating, noncorrosive check valve. Locate the check valve in the horizontal drain line (refer to *Figure 3*). The duplex arrangement allows wastewater to flow into the basin through one of the inlets while the other is closed during the pump discharge process. Install a pump alternator to switch between the two pumps. Occasionally, excess amounts of wastewater may surge into the basin, requiring both pumps to activate at the same time. To allow for this, install overflow strainers near the inlets for each pump. Overflow strainers allow wastewater to enter the sump when both check valves are closed.

2.1.2 Pneumatic Ejectors

A pneumatic ejector is an alternative to a centrifugal pump. Plumbers often select pneumatic ejectors for large capacity applications, such as housing subdivisions and municipal facilities. Unlike centrifugal pumps, pneumatic ejectors do not have moving parts. This means they require less maintenance than centrifugal pumps. However, they are more mechanically complex than centrifugal systems.

Sewage enters a pneumatic ejector through an inlet. When the wastes reach a predetermined height, a sensor triggers a compressor. The compressor forces air into the basin. The increased pressure causes the sewage to be moved into the discharge line and out of the basin. Pneumatic ejectors are equipped with check valves on the inlet and discharge lines. During waste ejection, the valves prevent back pressure. After ejection, they keep the waste from backflowing. When the ejector is empty, a diaphragm exhaust valve releases the compressed air and allows the chamber to refill with sewage (see *Figure 5*).

Pneumatic ejectors are designed as a unit. This means that basins, pumps, and controls are all part of a single manufactured package. Be sure to install the right size pneumatic ejector for the anticipated load. Consult the construction drawings, and check your local code for requirements governing pneumatic ejector installation. Install a union above the high-water line between the check valve and the pump to allow the pump to be removed for maintenance without disturbing the piping.

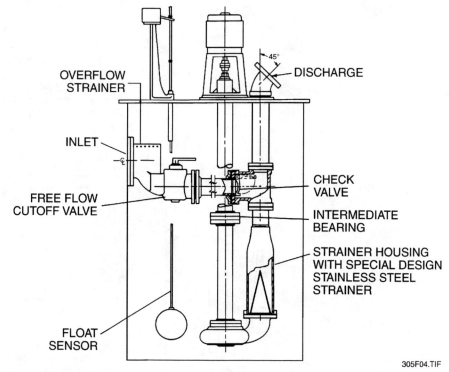

OVERFLOW
STRAINER

DISCHARGE

INLET

CHECK
VALVE

FREE FLOW
CUTOFF VALVE

INTERMEDIATE
BEARING

STRAINER HOUSING
WITH SPECIAL DESIGN
STAINLESS STEEL
STRAINER

FLOAT
SENSOR

305F04.TIF

Figure 4 ◆ Reverse flow pump installation.

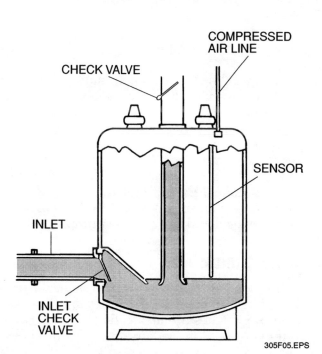

CHECK VALVE

COMPRESSED
AIR LINE

SENSOR

INLET

INLET
CHECK
VALVE

305F05.EPS

Figure 5 ◆ Pneumatic ejector.

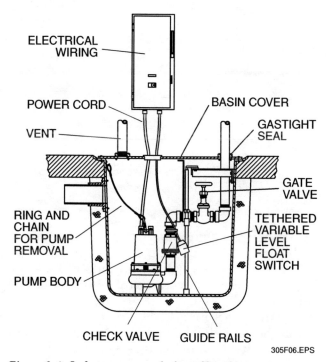

ELECTRICAL
WIRING

POWER CORD

VENT

BASIN COVER

GASTIGHT
SEAL

GATE
VALVE

RING AND
CHAIN
FOR PUMP
REMOVAL

TETHERED
VARIABLE
LEVEL
FLOAT
SWITCH

PUMP BODY

CHECK VALVE   GUIDE RAILS

305F06.EPS

Figure 6 ◆ Indoor sump made from fiberglass.

2.2.0 Sewage Sumps

A sump is a container that holds wastewater until it can be pumped into the sewer line. Plumbers are responsible for installing sumps. Installation includes all of the components within the sump, including pumps, controls, and inlet and outlet piping. Indoor sumps are made of plastic or fiberglass (see *Figure 6*). Outdoor sumps can be site-built from concrete (see *Figure 7*). Basin size and materials are specified in your local code. Install check valves on sump inlet piping, and install gate valves on discharge piping. Ensure that the valves are located outside the sump and are easily accessible. Residential installations may require only a check valve; refer to your local code.

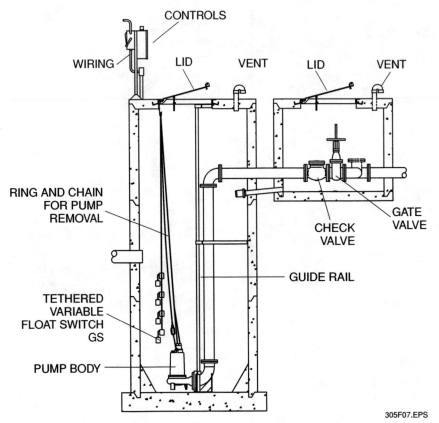

Figure 7 ◆ Outdoor sump basin made from concrete.

 WARNING!

To prevent ignition of sewer gas, always install explosion-proof pumps in sumps. Wear appropriate personal protective equipment when inspecting sewage pumps. Review the manufacturer's specifications before inspecting or removing a sewage pump.

When designing sumps, remember to consider the drain flow rate from all of the fixtures draining into the sewer line. Calculate the flow in terms of the **drainage fixture unit.** A drainage fixture unit is a measure of a fixture's discharge in gallons per minute divided by 7.5. This is the number of gallons in a cubic foot. Drainage fixture unit calculations take several factors into consideration:

- Rate of discharge
- Duration of a discharge
- Average time between discharges

Table 1 provides drainage fixture units for fixtures in commercial and private installations, and *Table 2* shows drainage fixture units for fixture drains or traps.

Install airtight covers and vents on sewage sumps. Otherwise, sewer gases will escape and

may cause illness and contamination. Remember that vents must always be separate from any required storm water vents. Consult local codes for specific venting requirements. Design the sewage removal system so that wastes are not retained for more than 12 hours. Maximum allowable height of wastewater in sumps is established in local codes. Ensure that the pump does not cycle more than six times per hour because excessive cycling could damage the pump.

2.3.0 Controls

When wastewater in a sump reaches a preselected height, a switch in the basin activates the pump. Switches are also called *sensors*. Sewage removal systems use any one of four types of switches:

- **Float switch**
- **Pressure float switch**
- **Mercury float switch**
- **Probe switch**

A float switch works a lot like the float in a water closet tank. As the wastewater level rises, the float rises within the sump. An arm attached to the float activates the pump when the float reaches a specified level within the sump.

Table 1 Drainage Fixture Units for Fixtures and Groups

Fixture Type	Drainage Fixture Unit Value as Load Factors	Minimum Size of Traps (inches)
Automatic clothes washers, commercial[1]	3	2
Automatic clothes washers, residential	2	2
Bathroom group (1.6 gpf water closet)[4]	5	—
Bathroom group (water closet flushing greater than 1.6 gpf)[4]	6	—
Bathtub[2] (with or without overhead shower or whirlpool attachment)	2	1½
Bidet	1	1¼
Combination sink and tray	2	1½
Dental lavatory	1	1¼
Dental unit or cuspidor	1	1½
Dishwashing machine,[3] domestic	2	1½
Drinking fountain	½	1½
Emergency floor drain	0	2
Floor drains	2	2
Kitchen sink, domestic	2	1½
Kitchen sink, domestic with food waster grinder and/or dishwasher	2	1½
Laundry tray (1 or 2 compartments)	2	1½
Lavatory	1	1¼
Shower	2	1½
Sink	2	1½
Urinal	4	*
Urinal, 1 gpf or less	2**	*
Water sink (circular or multiple) each set of faucets	2	1½
Water closet, flushometer tank, public or private	4**	*
Water closet, private (1.6 gpf)	3**	*
Water closet, private (flushing greater than 1.6 gpf)	4**	*
Water closet, public (1.6 gpf)	4**	*
Water closet, public (flushing greater than 1.6 gpf)	6**	*

[1] For traps larger than 3 inches, use Table 2.

[2] A showerhead over a bathtub or whirlpool bathtub attachment does not increase the drainage fixture unit value.

[3] See 2000 International Plumbing Code® Sections 709.2 through 709.4 for methods of computing unit value of fixtures not listed in this table or for rating devices with intermittent flow.

[4] For fixtures added to a dwelling unit bathroom group, add the DFU value of those additional fixtures to the bathroom group fixture count.

* Trap size shall be consistent with the fixture outlet size.

** For the purpose of computing loads on building drains and sewers, water closets or urinals shall not be rated at a lower draining fixture unit unless the lower values are confirmed by testing.

Table 2 Drainage Fixture Units for Fixture Drains or Traps

Fixture Drain or Trap Size (inches)	Drainage Fixture Unit Value
1¼	1
1½	2
2	3
2½	4
3	5
4	6

Source, Tables 1 and 2: International Code Council, Inc. See Figure and Table Credits, page 5.24.

A pressure float switch is an adjustable type of float switch. Install pressure float switches to rise with wastewater or clamp them to the basin wall (see *Figure 8*).

A mercury float switch is a specialized float sensor that plumbers can install at various heights within the basin (see *Figure 9*). Usually two mercury float switches are required, and a third switch can serve as a high water alarm if desired. If more than one pump is used in the basin, install four mercury float switches.

Probe switches, also called *electrodes*, are used in pneumatic ejectors. They work similarly to thermostatic probes installed on water heaters. When wastewater in the sump reaches a certain height, the probe triggers the compressor. When the waste is ejected, the switch resets.

Sewage removal systems perform a critical role in plumbing installations. Proper sizing and installation will help ensure that a system will not suffer from malfunctions and breakdowns. When possible, provide a backup pump in case the lead pump fails. For pneumatic ejector installations, provide a backup compressor. Power failures will prevent centrifugal pumps and pneumatic ejectors from working. If possible, provide a backup power supply. Remember that plumbers are responsible for helping to maintain public safety through the fast and efficient removal of wastes. These extra steps will allow the system to continue operating while plumbers are called in to make repairs.

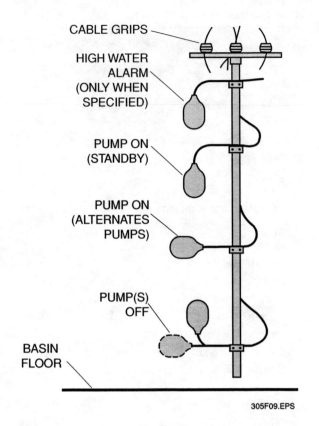

Figure 9 ◆ Mercury float switches in a sewage sump basin.

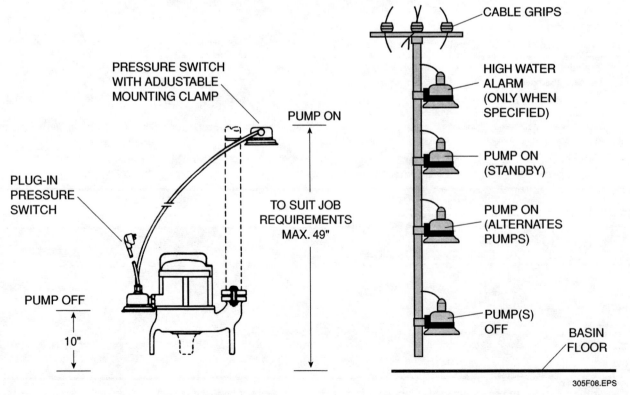

Figure 8 ◆ Pressure float switches in a sewage sump basin.

Review Questions

Sections 1.0.0–2.0.0

1. If it is not possible to install a check valve horizontally, install it _____.
   a. at a 45-degree angle
   b. vertically
   c. at a 60-degree angle
   d. parallel

2. To allow a pump to be removed without moving the piping, install a union above the high water line between the _____ and the pump.
   a. discharge pipe
   b. inlet pipe
   c. gate valve
   d. check valve

3. _____ allow wastewater to enter a sump when the check valves are closed.
   a. Gate valves
   b. Float switches
   c. Overflow strainers
   d. Cutoff valves

4. Calculate drainage fixture unit by dividing a fixture's discharge in gallons per minute by _____.
   a. .0133
   b. 3.14
   c. 7.5
   d. 8.3

5. When installing a mercury float switch, usually _____ switches are required.
   a. four
   b. two
   c. one
   d. six

3.0.0 ◆ STORM WATER REMOVAL SYSTEMS

Roofs, paved areas, and yards all require drains. The water from these areas drains into a separate storm sewer system or into special flood control areas, such as ponds and tanks. Many codes do not allow storm water to be discharged into sewage lines. Install a storm sewer line to drain clear water runoff. If the collection point for runoff must be deeper than the sewer drain line, install a storm water removal system to lift the wastewater.

Storm water removal systems are mainly used for flood control. The procedures for sizing and installing storm water removal systems share

DID YOU KNOW?

Pumps in History

Pumps have played a vital supporting role in some of history's greatest inventions and scientific discoveries. For example, in 1660, English chemist Robert Boyle (1627–1691) published a book called *The Spring and Weight of the Air* in which he used a hand pump of his own design to remove the air from a large sealed chamber. He placed a bell in the chamber and showed how the sound of the bell faded as more air was pumped out. He also placed a burning candle in the chamber and showed how it snuffed out as air was pumped out. His experiments with air pressure shattered many widely held beliefs about the nature of air. His work encouraged others to take up scientific experiments. Boyle's use of pumps to discover the nature of air is considered one of the most important examples of how empirical, or experimental, science can lead to new discoveries.

As a young man, the Scottish inventor and engineer James Watt (1736–1819) became interested in steam-powered pumps. Miners used primitive and inefficient steam-powered piston pumps to suck water out of England's coal mines. Watt did research to find ways to improve how the steam pumps worked. His research led him to discover the properties of steam, which in turn led to an improved steam engine design that eventually powered the Industrial Revolution in the nineteenth century.

The next time you install a sewage pump, remember that that humble pump has a noble pedigree!

some similarities with sewage removal systems. However, keep in mind that the applications of the two types of systems are very different. Storm water pumps and basins are designed for clear water wastes rather than wastes with solid matter. Local codes provide guidance on how to design storm water removal systems.

3.1.0 Storm Water Pumps

Pumps used in storm water removal systems are often called **sump pumps.** Sometimes they are also referred to as *bilge pumps.* Install centrifugal sump pumps to lift storm water from sumps to the drainage lines (see *Figure 10*). Like centrifugal sewage pumps, the motor of a centrifugal sump pump may be mounted either atop or inside the sump. Install pumps made of chlorinated polyvinyl chloride (CPVC) where the runoff contains corrosive chemicals (see *Figure 11*).

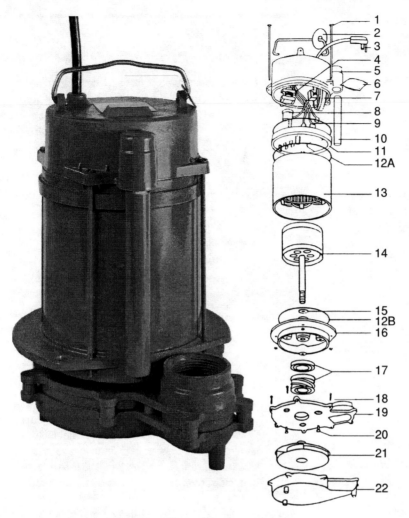

1	STAINLESS STEEL BOLTS
2	CARRYING HANDLE AND DATE TAG
3	POWER CORD
4	MOTOR START RELAY
5	MOTOR COVER
6	HUMISORB PACKET
7	WATER LEVEL SWITCH
8	OVERLOAD PROTECTOR
9	CONNECTING LEADS
10	PVC-AIR TUBE
11	BAFFLE AND TERMINAL PLATE
12A	RUBBER GASKET #050
12B	RUBBER GASKETS #160
13	STATOR ASSEMBLY — ½ HP — 115V 60C 7½ A CONTINUOUS DUTY OIL DUROL CC
14	ROTOR SHAFT ASSEMBLY
15	THRUST RACE
16	MOTOR END
17	CERAMIC SEAT AND ROTARY SEAL HEAD
18	STAINLESS STEEL SCREWS TYPE B
19	PUMP VOLUTE
20	FLAT HEAD SCREWS
21	REINFORCED POLYPROPYLENE IMPELLER
22	HOUSING PLATE

305F10.EPS

Figure 10 ◆ Submersible centrifugal sump pump.

PUMP MOTOR

DISCHARGE PIPE

PUMP BODY

INLET SCREEN

305F11.TIF

Figure 11 ◆ CPVC pump.

Many local codes require gate and check valves on the discharge line (see *Figure 12*). Check valves prevent backflow into the basin. Gate valves allow the plumber to shut off the discharge to allow maintenance. Aluminum flapper check valves are often specified because of their quieter operation. To reduce vibration, install a rubber base on the bottom of the pump and add a length of flexible hose in the discharge line.

Size the pump according to the anticipated amount of drainage. You will learn how to calculate rainfall rates later in this module. Provide a safety factor by ensuring that the pump has a capacity of 1¼ times the maximum calculated flow. An additional pump can serve as a backup in case the primary pump fails. Power failures often occur during storms, which is when sump pumps are most needed. Consider installing a backup power source to keep the sump pumps running.

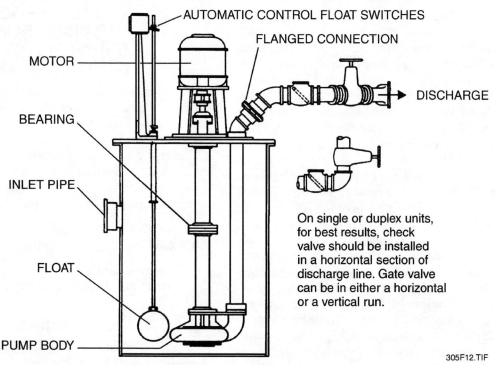

AUTOMATIC CONTROL FLOAT SWITCHES

FLANGED CONNECTION

MOTOR

DISCHARGE

BEARING

INLET PIPE

FLOAT

PUMP BODY

On single or duplex units, for best results, check valve should be installed in a horizontal section of discharge line. Gate valve can be in either a horizontal or a vertical run.

305F12.TIF

Figure 12 ◆ Check and gate valves in a centrifugal pump installation.

3.2.0 Storm Water Sumps

Storm water sumps collect runoff from sub-drains and store it until it can be pumped to storm sewers or to flood-control areas. Construct sumps from cast iron, fiberglass, or concrete. Apply waterproofing to the inside of concrete sumps. Provide a settling area inside sumps. This will allow sand and other coarse material to settle without being sucked up into the pump. Sediment can damage the pump. Sump covers do not need to be airtight; however, a secure lid will prevent objects from falling into the sump.

Sizing sumps properly will ensure that they perform their function efficiently. Sizing a storm water sump is similar to the process for sizing a sewage sump. Keep two objectives in mind when sizing a sump:

- The basin must be large enough to receive the anticipated amount of storm water.
- The basin must be large enough to eliminate frequent cycling of the pump motor.

Ensure that the basin's high water mark is at least 3 inches below the sub-drain inlet. The low-level mark must be no less than 6 inches above the bottom of the basin. This will ensure that the pump's suction end is always underwater.

Size the sump so that the pump motor does not activate more than once every 5 minutes. For example, if wastewater flows into a sump at 60 gallons per minute (gpm) and if the pump cycles every 5 minutes, then the basin has to hold at least 300 gallons (5 × 60) of wastewater.

How big will a sump have to be to handle 300 gallons of wastewater? You can size basins by using the formula for finding the volume of a cylinder. The formula is

$$V = (\pi R^2)h$$

In other words, the volume is the product of the area of the basin's floor and its height. You learned this formula earlier in *Plumbing Level Three*. When converting the volume to gallons, remember that there are 7.5 gallons in a cubic foot. Using the formula, a 4-foot diameter cylinder would have to be 3.18 feet tall, or 3 feet 2 inches, to hold 300 gallons. Adding the required 3-inch minimum at the top and 6-inch minimum at the bottom, the sump should be at least 3 feet 11 inches from the bottom of the inlet pipe to the base of the sump.

Local codes will provide the rate of wastewater collection and area rainfall rates to use when calculating the size of the removal system. Unless otherwise specified, use a collection rate of 2 gpm for every 100 square feet of sandy soil and 1 gpm for clay soils. Flow rates for roof and driveway runoff will vary considerably. For roofs and driveways, use a collection rate of 1 gpm for every 24 square feet of surface area. Local codes include specific guidelines for determining roof and driveway runoff rates.

To calculate the average amount of rain in the area, consult a table or map of local area rainfall. Tables and maps are printed in model and local codes, civil engineering handbooks, and landscape architecture manuals. You can also get information from the U.S. Department of Agriculture or the Weather Bureau of the U.S. Department of Commerce. The National Weather Service issues **100-year 1-hour rainfall maps** for the United States. These maps show the average recorded rainfall in 1 hour for a given area over the previous century. See *Appendix A* for an example of 100-year 1-hour rainfall maps for different parts of the country.

Design the removal system to discharge runoff into the building storm sewer. Local code may also allow the waste to discharge into a street gutter or curb drain. Some older installations connect storm drains with sewage waste lines. Many codes now prohibit this type of connection. Many cities and counties maintain **retention ponds** and **retention tanks** for storm water drainage. They are used for flood control. Retention ponds are large man-made basins where sediment and contaminants settle out of the wastewater before the water returns to rivers, streams, and lakes. Retention tanks are similar to retention ponds, except they are enclosed. Install backflow preventers on the drainage line to prevent contamination of the sump in case a retention pond or tank overflows.

3.3.0 Controls

Pump control is usually handled by a float sensor and/or a switch that activates the pump as the water level in the basin rises. Set the float to activate the pump at the highest water level possible inside the sump. This will prevent excessive cycling of the pump. Consider installing a standby pump or backup power system for emergency situations. An alarm will provide occupants with a warning signal during an emergency condition like a power outage. Review the installation requirements and construction drawings to see if these measures are appropriate.

 WARNING!

Review manufacturer's instructions before attempting to repair or replace a pump. Ensure that valves are closed and that electrical connections to the pump motor and controls have been turned off. Wear appropriate personal protective equipment. Do not work in a walk-in wastewater sump without prior training in how to work safely in confined spaces.

4.0.0 ◆ TROUBLESHOOTING AND REPAIRING SEWAGE AND STORM WATER REMOVAL SYSTEMS

Occasionally, pumps break down. When they do, they must be repaired or replaced. Plumbers can perform most pump repairs. A malfunctioning pump usually means an entire drainage system must be shut down until the pump can be repaired. Quick and efficient troubleshooting and repair or replacement is essential. Pump problems can usually be traced to one or both of the following causes:

- Electrical, involving the pump motor and wiring
- Mechanical, involving the pump's moving parts

The most common types of problems and their solutions are discussed below.

4.1.0 Troubleshooting Electrical Problems

Most problems with sewage and sump pumps can be traced to failure of the pump's electric motor. Power failures often happen during storms, when sump pumps are needed most. The loss of electrical power causes the pump to stop working, which puts the structure at risk of flooding. If the water level rises to the level of the pump motor, remove the electrical cord from the power source. If the motor is in danger of getting wet, it may be necessary to remove it from the pump. If a motor gets wet, ensure that it is thoroughly dry before connecting it to a power source. Consider installing a battery operated pump to operate the removal system during power outages.

If a pump is not running, check the power source to see if the voltage level is the same as the pump's required voltage. This is specified on the pump's nameplate. The problem may be a blown fuse or a broken or loose electrical connection. Be sure to check all fuses, circuit breakers, and electrical connections for breaks. If the power cord insulation is damaged, replace the cord with a new one of the same rating. If the power cord is open or grounded, you will need to contact an electrician to check the resistance between the cord's hot and neutral leads.

Occasionally, the impeller will lock. Have an electrician check the amps drawn by the pump motor. If the motor is drawing more amps than permitted by the manufacturer, the impeller may be blocked, the bearings may be frozen, or the impeller shaft may be bent. Remove the pump for inspection and repair or replace the damaged part. If a motor's overload protection has tripped, contact an electrician to inspect the motor.

Pumps sometimes fail to deliver their rated capacity because of electrical problems. If the impeller appears to be operating below normal speed, check the voltage supply against the motor's required voltage. Sometimes backflow can cause the impeller to rotate in the wrong direction. For single-phase electrical motors, simply shut off the power and allow the impeller to stop rotating. Then turn the pump on again; the problem should correct itself. For three-phase electrical motors, contact an electrician to inspect the motor (see *Figure 13*).

4.2.0 Troubleshooting Mechanical Problems

Nonelectrical, mechanical problems with pumps include worn, damaged, or broken controls or pump components. Float sensors are a common culprit in mechanical failures. Carefully inspect the floats for corrosion and sediment buildup,

which can cause floats to stick. To test if a float switch is broken, bypass the switch. If the pump operates, the switch will need to be replaced. Ensure the float is properly adjusted and that it has not slipped from its desired location. In any case, when repairing a waste removal system, always recalibrate and test the floats for proper operation.

Pneumatic ejectors use a rubber diaphragm exhaust valve to allow compressed air to escape and wastewater to resume filling the sump. If an ejector will not shut off, inspect the diaphragm and its switch. If the switch is broken or the rubber diaphragm is weak, replace the part immediately. Clean the areas around the diaphragm to remove any sediment lodged between the retainer ring and the diaphragm.

Plugged vent pipes can also prevent a pump from shutting off. Inspect the vent pipe and clear it out if necessary. The pump may have an **air lock**, which means that air trapped in the pump body is interfering with the flow of wastewater. Turn off the pump for about one minute, then restart it. Repeat this several times until the air is cleared from the pump. If the removal system has a check valve, ensure that a ³⁄₁₆-inch hole has been drilled in the discharge pipe about 2 inches above the connection. The wastewater inflow may match the pump's capacity. If this is the case, recalculate the amount of wastewater entering the system and install a suitable pump.

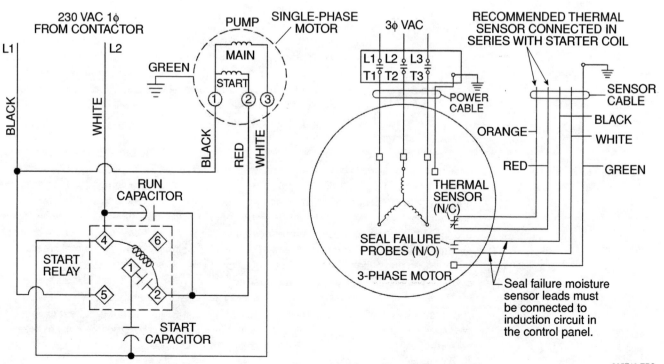

Figure 13 ◆ Single-phase and three-phase pump motors.

305F13.EPS

WARNING!
Storm water runoff may contain corrosive chemicals such as oil and grease. Wear appropriate personal protective equipment when performing maintenance on storm water sumps.

If the pump runs but won't discharge wastewater, the check valves may have been installed backwards. Inspect the flow-indicating arrow on the check valves. If the check valve is still suspect, remove it and test to make sure it is not stuck or plugged. Check the pump's rating to ensure the lift is not too high for the pump. Inspect the inlet to the impeller and remove any solids that may be blocking it. Turn the pump on and off several times to remove an air lock.

When a pump is not delivering its rated capacity, check to see if the impeller is rotating in the wrong direction. Wastewater can drain back into a sump from a long discharge pipe if a check valve is not installed. Reverse flow will cause the impeller to rotate backward. Turn the pump off until the impeller stops rotating. Then turn the pump back on. Install a check valve on the discharge line to prevent the problem from recurring. Inspect the impeller for wear due to abrasives or corrosion, and replace a worn impeller immediately. Finally, check the pump rating to ensure the lift is not too high for the pump.

If the pump cycles continually, the problem could again be wastewater draining back into the basin from the discharge line. Install a check valve in the discharge line. If a check valve is already installed, inspect it for leakage. Repair or replace leaking valves. The sump may be too small for the wastewater inflow. Recalculate the inflow and, if necessary, install a larger basin.

ON THE

Plumbing Code Requirements for Sewage and Storm Water Removal Systems

Local plumbing codes govern the installation and operation of sewage and storm water removal systems. Many local codes are based on one of the model plumbing codes. Always refer to your local code before installing sewage and storm water removal systems. Here are some general guidelines from two model codes: the International Plumbing Code® (IPC) and the Uniform Plumbing Code™ (UPC).

The UPC has two requirements for *sewage removal systems*: (1) the pump or discharge pipes connected to a water closet must be at least 2 inches in diameter; and (2) lines must have accessible backwater or swing check valves and gate valves.

The IPC's requirements for *sewage removal systems* are more specific. Lift stations must be covered with a gas-proof lid and must be vented. Dimensions of the basin must be at least 18 inches in diameter and 24 inches deep. Effluent (the system's outlet) must be kept at least 2 inches below the gravity drain system's inlet. For commercial installations, ejectors require a gate valve on the discharge side of the check valve before the gravity drain. Gate and check valves must be located above the lift station covers.

Additionally, the IPC mandates minimum ejector capacities: for discharge pipes 2 inches in diameter, the capacity must be 21 gpm; for discharge pipes 2½ inches in diameter, the capacity must by 30 gpm; and for discharge pipes 3 inches in diameter, the capacity must be 46 gpm.

According to the UPC, *storm water removal systems* must have sumps that are watertight and gravity-fed. The sumps can be made of steel, concrete, or other approved materials. Sumps must be vented, and the vent may be combined with other vent piping. In case of overload or mechanical failure, sumps located in "public use" areas must have dual pumps or ejectors that can function independently. Drain lines that connect to a horizontal gravity-fed discharge line must connect from the top through a wye. Gate valves are required on the discharge side of the backwater or gate valve.

The IPC's *storm water removal systems* call for sub-drains that must discharge into a sump serviced by an appropriately sized pump. Sub-soil sumps do not require gas-tight covers or vents. Sumps must be at least 18 inches in diameter. Electrical service outlets must conform to National Fire Protection Association (NFPA) standards. Discharge pipes require gate valve and full-flow check valve.

Odors associated with a sewage removal system may mean that the sump lid is not properly sealed. Check the cover to see if its seal is intact. Inspect the vent to ensure the sump has open-air access. Some commercial structures such as laundromats use special lint traps. If the sewer lines are backed up, check to make sure all the lint traps are clean.

4.3.0 Replacing Sewage and Storm Water Pumps

If a pump cannot be repaired, it will have to be replaced. When replacing a sewage or storm water pump, always remember to install a properly sized replacement. If a pump repeatedly breaks down, it may be the wrong size or type. Review the construction drawings, plumbing installation design, and removal system specifications. If necessary, have an engineer appraise the design and installation. These steps will ensure that the pump is matched with the job it is supposed to perform.

WARNING!

Review confined-space safety protocols before working in a walk-in sump. Wear appropriate personal protective equipment.

DID YOU KNOW?

The First Modern Sanitary Engineer

New York engineer Julius W. Adams laid the foundations for Brooklyn's sewer system in 1857. He ended up inventing a system that defined modern sanitary engineering. His design provided sewage drains for all of Brooklyn's 20 square miles. Adams used technology already in use by other cities—reservoirs feeding into underground mains that branched off to connect commercial and residential buildings. However, Adams did something that no one had done before: he distilled his experience into a sewer design manual, which he then published. For the first time, cities throughout the country could consult a standard reference for sewer design.

Many pumps are water-lubricated. This means that they must be immersed in water before they can be tested or run. Pumps are often self-priming. This means they automatically fill with water before operating. Failure to run the pump in water could burn out the pump's bearings quickly. Follow manufacturer's specifications closely when installing and testing sump and sewage pumps.

Review Questions

Sections 3.0.0–4.0.0

1. When sizing a sump pump, ensure that it has a capacity of _____ times the maximum calculated flow.
   a. 1¼
   b. 2
   c. 2½
   d. 3

2. The formula for determining the volume of a sump is _____.
   a. $V = (\pi R2)h$
   b. $V = lwh$
   c. $V = \frac{1}{2}(lwh)$
   d. $V = \pi R2$

3. Install float sensors to activate the pump at the _____ water level in the basin.
   a. lowest possible
   b. average
   c. highest possible
   d. initial

4. Most pump problems can be traced to failure of the _____.
   a. impeller
   b. switch
   c. power cord
   d. motor

5. _____ can cause an impeller to rotate in the wrong direction.
   a. Air lock
   b. Backflow
   c. Weak seals
   d. Corrosion

5.0.0 ◆ THE WORKSHEET

Refer to the appropriate sections in the module to answer the following questions.

1. What are the two types of pump that can be used in sewage removal systems?

   _____

2. How does an impeller work?

   _____

3. What are the reasons for installing a duplex pump?

   _____

4. What three considerations should a plumber keep in mind when calculating drainage fixture unit?

   _____

5. Refer to *Appendix B*. Identify the various parts of a sewage waste removal system pointed out in the illustration.

   _____

6. List four different types of pump switches.

   _____

7. What collection rate can plumbers use to estimate runoff from roofs and driveways?

   _____

8. Name at least two causes of impeller lock.

   _____

9. Describe the procedure for inspecting a suspect float.

   _____

10. What are the mechanical reasons a pump will cycle constantly?

    _____

Summary

Sewage and storm water removal systems are important components of plumbing installation. They pump wastewater in sub-drains into the sewer line where they can discharge into a public sewer, retention pond, or retention tank. Pneumatic ejectors and centrifugal pumps are used to lift wastewater from basins into the discharge line. Sumps hold the wastewater until the pump discharges it. Switches installed on the inside of the basin trigger the pump when the wastewater reaches a preselected height.

Plumbers are responsible for maintaining, repairing, and replacing pumps. Electrical and mechanical problems can be diagnosed by inspecting various components. Always remember to take appropriate precautions when working with electrical pumps near water. Consult with engineers and electricians if the problem requires more than the repair or replacement of a part or a whole pump. Because sewage and storm water removal systems help maintain public health and safety, respond to pump problems quickly and professionally.

Trade Terms Introduced in This Module

100-year 1-hour rainfall map: A chart showing an area's average recorded rainfall, occurring in 1 hour, over the previous century.

Air lock: A pump malfunction caused by air trapped in the pump body, interrupting wastewater flow.

Centrifugal pump: A device that uses an *impeller* powered by an electric motor to draw wastewater out of a *sump* and discharge it into a drain line.

Drainage fixture unit: A measure of the discharge in gallons per minute of a fixture, divided by the number of gallons in a cubic foot.

Duplex pump: Two *centrifugal pumps* and their related equipment installed in parallel in a *sump.*

Ejector: A common alternative name for a *sewage pump.*

Float switch: A container filled with a gas or liquid that measures wastewater height in a sump and activates the pump when the float reaches a specified height.

Impeller: A set of vanes or blades that draws wastewater from a *sump* and ejects it into the discharge line.

Lift station: A popular name for a *sewage removal system.* The term is also used to refer to a pumping station on a main sewer line.

Mercury float switch: A *float switch* filled with mercury. It activates the pump when it reaches a specified height in a basin.

Pneumatic ejector: A pump that uses compressed air to force wastewater from a *sump* into the drain line.

Pressure float switch: An adjustable float switch that can either be allowed to rise with wastewater or be clipped to the *sump* wall.

Probe switch: An electrical switch used in *pneumatic ejectors* that closes on contact with wastewater.

Retention pond: A large man-made basin into which storm water runoff drains until sediment and contaminants have settled out.

Retention tank: An enclosed container that stores storm water runoff until sediment and contaminants have settled out.

Reverse flow pump: A *duplex pump* designed to discharge solid waste without allowing it to come into contact with the pumps' *impellers.*

Sewage pump: A device that draws wastewater from a *sump* and pumps it to the sewer line. It may be either a *centrifugal pump* or a *pneumatic ejector.*

Sewage removal system: An installation consisting of a *sump*, pump, and related controls that stores sewage wastewater from sub-drains and lifts it into the main drain line.

Storm water removal system: An installation consisting of a *sump*, pump, and related controls that stores storm water runoff from a *sub-drain* and lifts it into the main drain line.

Sub-drain: A building drain located below the sewer line.

Sump: A container that collects and stores wastewater from a *sub-drain* until it can be pumped into the main drain line.

Sump pump: A common name for a pump used in a *storm water removal system.*

100-Year, 1-Hour Rainfall in Inches

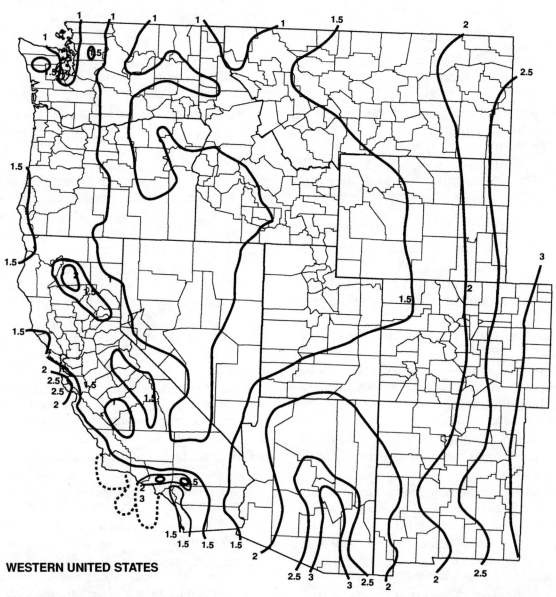

WESTERN UNITED STATES

FOR SI: 1 INCH = 25.4 MM.
SOURCE: NATIONAL WEATHER SERVICE, NATIONAL OCEANIC AND ATMOSPHERIC ADMINISTRATION, WASHINGTON, DC.

305A01.EPS

Source: International Code Council, Inc. See Figure and Table Credits, page 5.24.

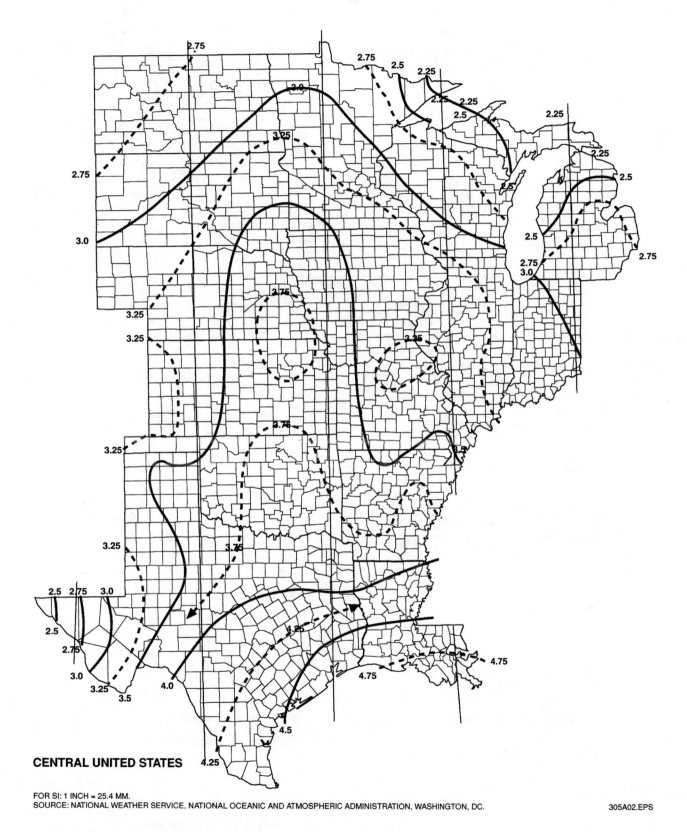

CENTRAL UNITED STATES

FOR SI: 1 INCH = 25.4 MM.
SOURCE: NATIONAL WEATHER SERVICE, NATIONAL OCEANIC AND ATMOSPHERIC ADMINISTRATION, WASHINGTON, DC.

305A02.EPS

Source: International Code Council, Inc. See Figure and Table Credits, page 5.24.

EASTERN UNITED STATES

FOR SI: 1 INCH = 25.4 MM.
SOURCE: NATIONAL WEATHER SERVICE, NATIONAL OCEANIC AND ATMOSPHERIC ADMINISTRATION, WASHINGTON, DC.

305A03.EPS

Source: International Code Council, Inc. See Figure and Table Credits, page 5.24.

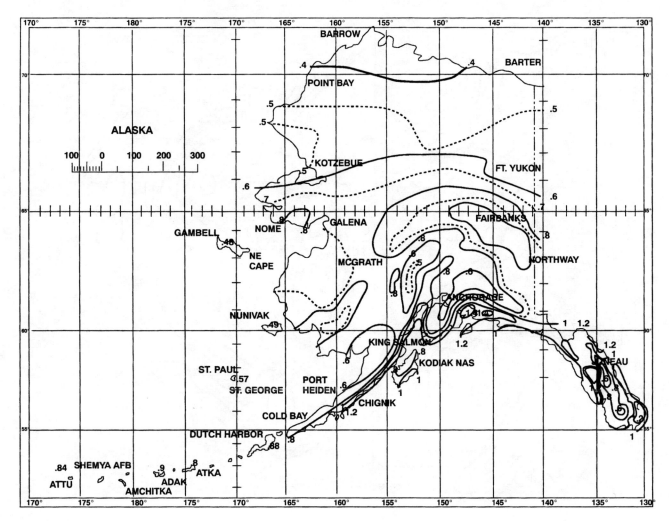

ALASKA

FOR SI: 1 INCH = 25.4 MM.
SOURCE: NATIONAL WEATHER SERVICE, NATIONAL OCEANIC AND ATMOSPHERIC ADMINISTRATION, WASHINGTON, DC.

305FA04.EPS

Source: International Code Council, Inc. See Figure and Table Credits, page 5.24.

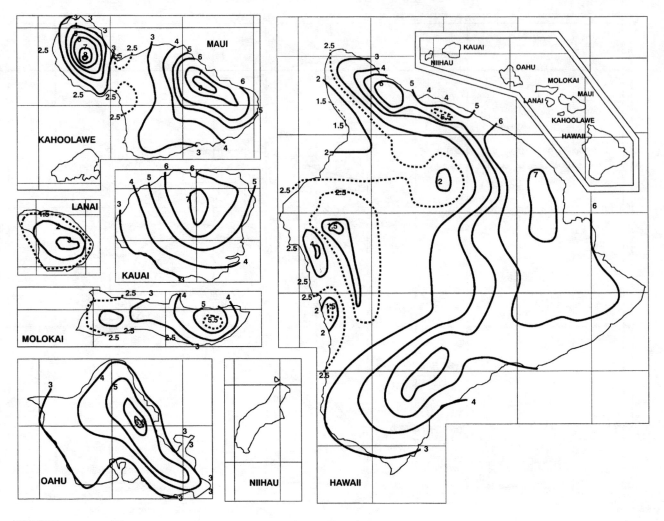

HAWAII

FOR SI: 1 INCH = 25.4 MM.
SOURCE: NATIONAL WEATHER SERVICE, NATIONAL OCEANIC AND ATMOSPHERIC ADMINISTRATION, WASHINGTON, DC.

305A05.EPS

Source: International Code Council, Inc. See Figure and Table Credits, page 5.24.

Worksheet Illustration

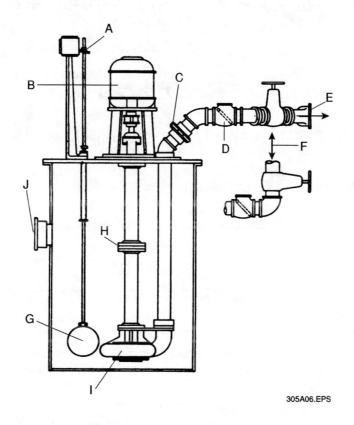

305A06.EPS

Additional Resources

This module is intended to present thorough resources for task training. The following reference works are suggested for further study. These are optional materials for continued education rather than for task training.

Modern Plumbing, 1997. E. Keith Blankenbaker. Tinley Park, IL: Goodheart-Willcox Company.

Planning Drain, Waste & Vent Systems, 1993. Howard C. Massey. Carlsbad, CA: Craftsman Book Company.

References

Planning Drain, Waste & Vent Systems, 1993. Howard C. Massey. Carlsbad, CA: Craftsman Book Company.

1997 International Plumbing Code, 1997. Falls Church, VA: International Code Council.

Building Services Engineering: A Review of Its Development, 1982. Neville S. Billington. New York: Pergamon Press.

Figure and Table Credits

Tables 1 and 2 and Appendix A figures: Written permission to reproduce this material was sought from, and granted by, the copyright holder, **International Code Council, Inc.,** 5203 Leesburg Pike, Suite 600, Falls Church, VA 22041.

Stevens Pump Company	305F01, 305F10
Zoeller Pump Company	305F06, 305F07, 305F13

NCCER CRAFT TRAINING USER UPDATES

The NCCER makes every effort to keep these textbooks up-to-date and free of technical errors. We appreciate your help in this process. If you have an idea for improving this textbook, or if you find an error, a typographical mistake, or an inaccuracy in the NCCER's Craft Training textbooks, please write us, using this form or a photocopy. Be sure to include the exact module number, page number, a detailed description, and the correction, if applicable. Your input will be brought to the attention of the Technical Review Committee. Thank you for your assistance.

Instructors – If you found that additional materials were necessary in order to teach this module effectively, please let us know so that we may include them in the Equipment and Materials list in the Instructor's Guide.

Write: Curriculum Revision and Development Department
National Center for Construction Education and Research
P.O. Box 141104, Gainesville, FL 32614-1104

Fax: 352-334-0932

E-mail: curriculum@nccer.org

Craft _____   Module Name _____

Copyright Date _____   Module Number _____   Page Number(s) _____

Description _____

_____

_____

_____

(Optional) Correction _____

_____

_____

(Optional) Your Name and Address _____

_____

_____

_____

Sizing Water Supply Piping

COURSE MAP

This course map shows all of the modules in the third level of the Plumbing curriculum. The suggested training order begins at the bottom and proceeds up. Skill levels increase as you advance on the course map. The local Training Program Sponsor may adjust the training order.

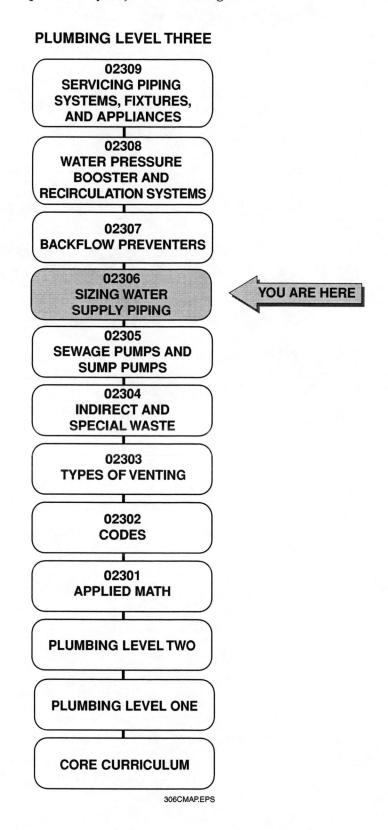

PLUMBING LEVEL THREE

02309
SERVICING PIPING
SYSTEMS, FIXTURES,
AND APPLIANCES

02308
WATER PRESSURE
BOOSTER AND
RECIRCULATION SYSTEMS

02307
BACKFLOW PREVENTERS

02306
SIZING WATER
SUPPLY PIPING
◁ YOU ARE HERE

02305
SEWAGE PUMPS AND
SUMP PUMPS

02304
INDIRECT AND
SPECIAL WASTE

02303
TYPES OF VENTING

02302
CODES

02301
APPLIED MATH

PLUMBING LEVEL TWO

PLUMBING LEVEL ONE

CORE CURRICULUM

306CMAP.EPS

MODULE 02306 CONTENTS

Figures

Tables

Sizing Water Supply Piping

Objectives

When you have completed this module, you will be able to do the following tasks in accordance with local codes:

1. Calculate pressure drops in a water supply system.
2. Size pipe for different flow rates.
3. Explain the difference between and advantages of a continuous-flow system and an intermittent-flow system.
4. Identify fixtures with high flow rates.
5. Explain the proper viscosity of liquids used in water supply installation.
6. Lay out a water supply system.
7. Calculate developed lengths of branches for a given water supply system.
8. Calculate flow rates for high flow rate fixtures.

Prerequisites

Before you begin this module, it is recommended that you successfully complete the following: Core Curriculum; Plumbing Level One; Plumbing Level Two; Plumbing Level Three, Modules 02301 through 02305.

Required Trainee Materials

1. Appropriate personal protective equipment
2. Pencil and paper
3. Copy of your local code

1.0.0 ◆ INTRODUCTION

People take their fresh water supply for granted. No one thinks twice when water comes out of a spigot when they turn on a faucet. However, they *do* notice when they turn the handle and nothing comes out. The water supply system is one of the most important installations that plumbers put in. It is a vital part of residential and commercial buildings. Sizing water supply systems is a complex but important task. Supply systems need to provide adequate water at the correct pressure, and they need to do this reliably and efficiently every time.

You learned how to install water supply piping in *Plumbing Level Two*. This module discusses the general concepts used for sizing water supply systems. You will learn how to calculate a system's requirements and **demand.** When you are finished, you will be able to size a water supply system so that it provides the right amount of water, at the right pressure, at the right time.

2.0.0 ◆ FACTORS AFFECTING WATER SUPPLY PIPING

To install water supply systems correctly, you will need to understand the physical properties of water. These properties affect how water behaves inside pipes and fittings. Plumbers must consider the following factors when sizing a water supply system:

- Temperature
- **Density**
- Flow
- **Friction**

Each of these factors has an effect on the operation of a water supply system. It is important to install water supply systems so that the effects of those factors are limited. This will help the system last longer and operate more efficiently.

2.1.0 Temperature and Density

You have already learned about the importance of temperature and of one of its related properties, pressure. Another property of temperature that affects plumbing systems is density. Density is the amount of a substance in a given space. It is measured in pounds per cubic foot. Increasing the temperature of a substance decreases its density (see *Figure 1*). It also increases its volume. Plumbers need to know the temperature of water in a plumbing system to choose the right size pipe.

Another way to explain density is to say that as a volume of water gets hotter, it gets lighter. At 32° F, a cubic foot of water weighs 62.42 pounds, but at 100°F, the same amount of water weighs 61.99 pounds. At 200°F, it weighs 60.14 pounds. This means that hot water needs a bigger pipe to achieve the same amount of flow as cold water. The temperature of the water helps determine what diameter the supply pipe needs to be to handle the amount of water required by the system.

Overheated water is dangerous. It can split pipes and weaken fittings. It can also cause water heaters to explode. Install temperature/pressure relief valves on hot water heaters to prevent overheated water from causing damage (see *Figure 2*). Relief valves release water or steam and equalize the pressure in the tank. You learned how to install relief valves in *Plumbing Level Two*. Review your local code to determine the proper precautions for hot water pipes.

CHANGES IN WATER DENSITY AS A FUNCTION OF HEAT
(DATA BASED ON A CUBIC FOOT OF WATER AT SEA LEVEL)

306F01.EPS

Figure 1 ◆ The density of a cubic foot of water at different temperatures.

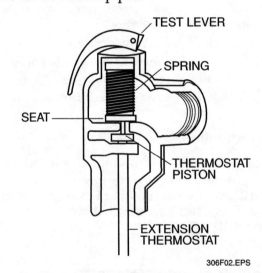

306F02.EPS

Figure 2 ◆ A temperature/pressure relief valve.

2.2.0 Flow

In plumbing, flow is the measure of how much liquid moves through a pipe. The measure of how much a liquid resists flow is called **viscosity.** You may have encountered this term in reference to motor oil. High-viscosity oil is very thick. In other words, it resists flowing. The concepts of flow and resistance are very important in water supply systems. There are three types of flow:

- **Laminar flow**
- **Transient flow**
- **Turbulent flow**

In cases where the rate of flow is less than one foot per second, water flows smoothly. If you could see the molecules in slow-moving water, they would look like they were moving parallel to each other in layers. This is called *laminar flow.*

Laminar flow is also called *streamline flow* or *viscous flow.* When water flow changes from one type of flow to another, it becomes erratic and unstable. This type of flow is called *transient flow.*

Turbulent flow is the random motion of water as it moves along a rough surface, such as the inner surface of a pipe. The faster the water flows along the inner surface, the more turbulent the flow becomes. Plumbers are mostly concerned with turbulent flow.

The different types of flow can be visualized by thinking about how water behaves in a kitchen sink. The water coming out of the faucet moves in a straight line. This is laminar flow. In the basin, it sloshes and spins. This is turbulent flow. The moment at which the water hits the basin and starts to swirl—you may not even be able to see this—is transient flow.

DID YOU KNOW?

Leonardo da Vinci and Turbulent Flow

Why does water flow the way it does? The famous Italian Renaissance scientist and artist Leonardo da Vinci (1452–1519) may have been the first to look for the answer. Da Vinci loved to study nature. He made sketches showing how different parts of the human body worked. He studied how animals moved. Eventually, he turned his attention to the motion of water.

Da Vinci carefully studied the way water flowed into a basin. He placed objects in the stream and watched how the water flowed around and over them. He made many detailed sketches of the patterns he saw. He was the first to use the word "turbulence" ("turbolenza") to describe the patterns. He claimed that turbulent flow consisted of three phases. The first phase happens when water hits an obstacle. Next, as the water flows around the obstacle, it creates whirls and eddies. Finally, the water flow becomes smooth again as it travels farther from the obstacle.

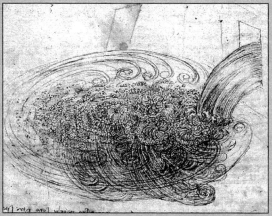

306SA01.EPS

Detail from "Notebook of Leonardo, known today as the Codex Arundel," Manuscript MS 263, ff 160b–161, The Royal Collection © 2001, Her Majesty Queen Elizabeth II.

Scientists were not able to describe turbulent flow using mathematics until the 19th century. Even today, scientists do not completely understand the process. Today's scientists use powerful computers to analyze how water flows, but the basic principles that Leonardo da Vinci first described 500 years ago still stand.

2.3.0 Friction

Turbulent flow is caused by friction. Friction is the resistance or slowing down that happens when two things rub against each other. For example, run your hand over a smooth surface, like silk or a polished tabletop. Your hand moves easily over the surface. It is harder to run your hand over a bumpy rug or over sandpaper, because the uneven surface slows your hand down. This is an example of friction.

Friction makes water supply systems less efficient. It reduces the pressure in the system by slowing water down. This is called **friction loss**. Friction loss is also known as *pressure loss*. In water supply piping, friction is caused by several factors, including:

- The inner surface of a pipe
- Flow through water meters
- Flow through fittings such as valves, elbows, and tees
- Increased water velocity

Effective system installation can reduce overall friction and improve system efficiency. Plumbers can reduce friction caused by pipes by increasing the pipe size. This is especially true when the water pipe is undersized. When water meters are required, use one that is designed for low friction. Keep the number of valves to a minimum to prevent excessive flow restriction. See the section on sizing water supply piping in this module to learn how to calculate friction loss in the system.

There is much less friction at the center of a pipe than near its inner surface. This means that the **flow rate,** or the speed that water flows, is faster at the center of a pipe and slower near the pipe's inner surface (see *Figure 3*). Flow rate is measured in gallons per minute (gpm). Plumbers install plumbing systems to minimize turbulence and maximize flow. Plumbers can consult published sources to determine the average rate of flow under different conditions. The average rate of flow is determined by a combination of available pressure, demand, system size, and pipe size.

Fittings and valves can also cause friction loss. Calculate the amount of friction loss caused by fittings and valves by comparing them with lengths of pipe that would cause equal friction (see *Table 1*). These are called **equivalent lengths.** You can use equivalents to calculate **pressure drop.** This is the difference in pressure between the inlet and the farthest outlet. You will learn more about pressure drop later in this module.

Refer to the table to determine the most efficient choice of fittings. For example, if you are going to install two 45-degree elbows to make a right-angle

bend and the system uses ⅝-inch tubing, consult the table. One 45-degree elbow causes the same friction loss as 0.5 feet of pipe. Therefore, the friction losses in two 45-degree elbows equal the loss in 1 foot of pipe. However, one 90-degree elbow causes the same friction loss as 1.5 feet of pipe. Therefore, if the length of pipe between the two 45-degree elbows was less than 0.5 foot, the use of two 45-degree elbows would be the more efficient choice (see *Figure 4*). Be sure to refer to local code for standards in your area.

The table also allows you to compare the friction losses caused by various types of valves. A rule of thumb that many plumbers use is to allow an additional 50 percent of the system's total length for fittings and valves. You will learn how to calculate the total length later in this module. Remember to test the accuracy of your estimate after the pipe sizes have been determined.

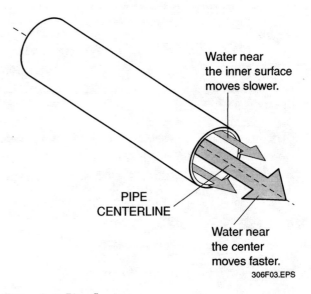

Figure 3 ◆ Pipe flow rates.

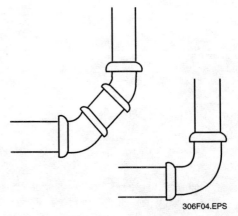

Figure 4 ◆ Friction loss comparison between two 45-degree elbows and one 90-degree elbow.

Table 1 Sample Equivalent Pipe Lengths Used for Determining Friction Loss in Fittings

	Friction Loss in Fittings and Valves Expressed as an Equivalent Length of Tube, in Feet								
Nominal or Standard Size (inches)	**FITTINGS**				**VALVES**				
	Standard Ell		**90-Degree Tee**						
	90-Degree	**45-Degree**	**Side Branch**	**Straight Run**	**Coupling**	**Ball**	**Gate**	**Butterfly**	**Check**
⅜	0.5	—	1.5	—	—	—	—	—	1.5
½	1.0	0.5	2.0	—	—	—	—	—	2.0
⅝	1.5	0.5	2.0	—	—	—	—	—	2.5
¾	2.0	0.5	3.0	—	—	—	—	—	3.0
1	2.5	1.0	4.5	—	—	0.5	—	—	4.5
1¼	3.0	1.0	5.5	0.5	0.5	0.5	—	—	5.5
1½	4.0	1.5	7.0	0.5	0.5	0.5	—	—	6.5
2	5.5	2.0	9.0	0.5	0.5	0.5	0.5	7.5	9.0
2½	7.0	2.5	12.0	0.5	0.5	—	1.0	10.0	11.5
3	9.0	3.5	15.0	1.0	1.0	—	1.5	15.5	14.5
3½	9.0	3.5	14.0	1.0	1.0	—	2.0	—	12.5
4	12.5	5.0	21.0	1.0	1.0	—	2.0	16.0	18.5
5	16.0	6.0	27.0	1.5	1.5	—	3.0	11.5	23.5
6	19.0	7.0	34.0	2.0	2.0	—	3.5	13.5	26.5
8	29.0	11.0	50.0	3.0	3.0	—	5.0	12.5	39.0

Source: International Code Council, Inc. See Figure and Table Credits, page 6.18.

Review Questions

Sections 1.0.0–2.0.0

1. When sizing a water supply system, plumbers must consider temperature, density, flow, and ___*a*___ .

   a. friction
   b. turbulence
   c. pressure
   d. demand

2. Increasing the temperature of a substance ___*d*___ its density.

   a. does not alter
   b. equalizes
   c. increases
   d. decreases

3. Of the three types of flow, plumbers are most concerned with ___*C*___ .

   a. laminar flow
   b. viscous flow
   c. turbulent flow
   d. transient flow

4. A viscous liquid tends to flow more ___*b*___ than a nonviscous liquid.

   a. slowly
   b. quickly
   c. freely
   d. distance

5. Increasing the velocity of water in a pipe ___*a*___ friction.

   a. decreases
   b. increases
   c. initiates
   d. does not affect

3.0.0 ◆ LAYING OUT THE WATER SUPPLY SYSTEM

Plumbers make sure that water supply systems work. Although systems differ from building to building, they are all built using the same basic rules. Plumbers must be able to determine the water requirements for all of the fixtures and outlets in the system. They must calculate the total length of pipes and fittings and how much water they can handle. All this information helps plumbers install a system that will meet the needs of the building's occupants. In this section, you will learn how to determine the water requirements and demand in a water supply system.

CAUTION

Water hammer can reduce the life of the components of a water supply system. Install a water hammer arrestor near each quick-closing valve in the water supply system.

3.1.0 System Requirements

Plumbers must consider a number of factors when installing a water supply system. Those factors include the following:

- The system's ability to carry an adequate supply of water.
- The system's ability to operate with a minimum of turbulence.
- The system's ability to maintain a reasonable flow rate, which will ensure quiet operation.
- The system's design—it should not be expensively overdesigned.

Careful installation will ensure that the system provides the best possible service. Before you begin, consult the building's plans or blueprints. Plumbing drawings are included in the plans for most industrial and commercial buildings. For light residential buildings, consult the floor plan. The drawings will help you understand how the system should be laid out.

Refer to the material takeoff for the types and number of fixtures and outlets used in the system. Determine the flow rate at each fixture and outlet (see *Table 2*). The flow rate is the measure of water flow required by the fixture to operate. It is calculated in gallons per minute (gpm). This information can also be found in the product specifications. Identify both the rate and pressure of flow for each item. Note that the rate and pressure may be less if water conservation devices are used. Determining the flow rate will provide an initial idea of the size of the system.

Table 2 Demand and Flow Pressure Values for Fixtures

Fixture Supply Outlet Serving	Flow Rate (gpm)	Flow Pressure (psi)
Bathtub	4.00	8
Bidet	2.00	4
Combination fixture	4.00	8
Dishwasher, residential	2.75	8
Drinking fountain	0.75	8
Laundry tray	4.00	8
Lavatory	2.00	8
Shower	3.00	8
Shower, temperature controlled	3.00	20
Sill cock, hose bibb	5.00	8
Sink, residential	2.50	8
Sink, service	3.00	8
Urinal, valve	15.00	15

Source: International Code Council, Inc. See Figure and Table Credits, page 6.18.

ON THE LEVEL

Sometimes the water pressure serving a building is too low for the system's requirements. In these cases, codes require that water pressure booster systems be installed. Fixture outlet pressures vary from code to code. The IPC, for example, requires a flow pressure of 8 pounds per square inch (psi) for most common household fixtures. A temperature-controlled shower requires up to 20 psi, and some urinals require 15 psi. Check your local code's pressure requirements before installing the system.

One of the goals of an efficient system installation is to supply adequate water pressure to the farthest point of use during peak demand. If the system can do that, then the points in between are also likely to have adequate service. However, this assumption may not always be true. Pressure and water volume requirements will vary from fixture to fixture. Keep these issues in mind when installing the water supply system.

Some fixtures may have a recommended pressure that is lower than the system's pressure. To correct this, install a flow restrictor on the line before the fixture. Flow restrictors can also reduce the system's water and energy consumption. This is especially true for hot water lines. In a high-rise building, pressure at one floor may exceed that at another. Install pressure-reducing valves to balance the pressure throughout the system.

It is also important to maintain the correct pressure within the system. Too little pressure will result in poor service. Negative pressure can cause back siphonage, which could contaminate the system. Excess pressure can blow out fixture traps. It can also cause the system to emit an objectionable whistling noise. Pipes that are too small can also cause whistling.

Some fixtures might require a larger volume of water than others. The installation will have to include a larger pipe to supply that fixture. Remember that water flows more slowly through a wide pipe than a narrow one. Ensure that the rate of flow does not fall below the fixture's requirements.

ON THE · LEVEL ·

Standards Organizations in the IPC

Plumbers using codes based on the International Plumbing Code® (IPC) must use pipes and fittings that conform to strict standards. These standards have been developed by several organizations. Two of the most frequently cited are located in the United States. Another is in Canada.

Take the time to obtain and read these standards carefully. Talk to local plumbing experts and learn how the standards are used in your area. This information will help you broaden your professional knowledge. It will also be reflected in the quality of your work.

American Society for Testing and Materials (ASTM)
100 Barr Harbor Drive
West Conshohocken, PA 19428-2959
www.astm.org

ASTM develops voluntary consensus standards and related technical information. It provides public health and safety guidelines, reliability standards for materials and services, and standards for commerce.

Canadian Standards Association (CAN/CSA)
178 Boulevard Rexdale
Toronto, Ontario
M9W 1R3, Canada
www.csa.ca

CSA develops standards for public safety and health, the environment, and trade. It also provides education and training for people who use their standards.

American Water Works Association (AWWA)
6666 W. Quincy Ave.
Denver, CO 80235
www.awwa.org

AWWA is a not-for-profit organization dedicated to improving the quality and supply of drinking water. It specializes in scientific research and educational outreach. AWWA is the world's largest organization of water supply professionals.

3.2.0 Calculating Demand

After determining the system's requirements, estimate its demand. The demand is the water requirement for the entire system—pipes, fittings, outlets, and fixtures. Plumbers calculate the rate of flow in gpm according to the number of **water supply fixture units** (WSFUs) that each fixture is designed to handle. WSFUs measure a fixture's load. WSFUs vary with the quantity and temperature of the water and with the type of fixture (see *Table 3*). Determine the WSFUs for all fixtures and outlets in the system. Then convert the WSFUs to gpm (see *Table 4*). Add up the results for all the fixtures and outlets. This is the total capacity of all the fixtures in the water supply system.

Table 3 Load Values Assigned to Fixtures

| Fixture | Occupancy | Type of Supply Control | Load Values in WSFUs | | |
			Cold Water	Hot Water	Total Load
Bathroom group	Private	Flush tank	2.70	1.50	3.60
Bathroom group	Private	Flush valve	6.00	3.00	8.00
Bathtub	Private	Faucet	1.00	1.00	1.40
Bathtub	Public	Faucet	3.00	3.00	4.00
Bidet	Private	Faucet	1.50	1.50	2.00
Combination fixture	Private	Faucet	2.25	2.25	3.00
Dishwashing machine	Private	Automatic	—	1.40	1.40
Drinking fountain	Offices, etc.	⅜-inch valve	0.25	—	0.25
Kitchen sink	Private	Faucet	1.00	1.00	1.40
Kitchen sink	Hotel, Restaurant	Faucet	3.00	3.00	4.00
Laundry trays (1–3)	Private	Faucet	1.00	1.00	1.40
Lavatory	Private	Faucet	0.50	0.50	0.70
Lavatory	Public	Faucet	1.50	1.50	2.00
Service sink	Offices, etc.	Faucet	2.25	2.25	3.00
Shower head	Public	Mixing valve	3.00	3.00	4.00
Shower stall	Private	Mixing valve	1.00	1.00	1.40
Urinal	Public	1-inch Flush valve	10.00	—	10.00
Urinal	Public	¾-inch Flush valve	5.00	—	5.00
Urinal	Public	Flush tank	3.00	—	3.00
Washing machine (8 lbs.)	Private	Automatic	1.00	1.00	1.40
Washing machine (8 lbs.)	Public	Automatic	2.25	2.25	3.00
Washing machine (15 lbs.)	Public	Automatic	3.00	3.00	4.00
Water closet	Private	Flush valve	6.00	—	6.00
Water closet	Private	Flush tank	2.20	—	2.20
Water closet	Public	Flush valve	10.00	—	10.00
Water closet	Public	Flush tank	5.00	—	5.00
Water closet	Public or Private	Flushometer tank	2.00	—	2.00

Note: For fixtures not listed, loads should be assumed by comparing the fixture to a listed fixture that uses water in similar quantities and at similar rates. The assigned loads for fixtures with both hot and cold water supplies are given for separate hot and cold water loads and for total load, where the separate hot and cold water loads are three-fourths of the total for the fixture in each case.

Source: International Code Council, Inc. See Figure and Table Credits, page 6.18.

Table 4 Table for Estimating Demand

Supply System Predominantly for Flush Tanks			Supply System Predominantly for Flush Valves		
Load	Demand		Load	Demand	
Water Supply Fixture Units	gpm (gallons/min.)	cfm (cubic ft./min.)	Water Supply Fixture Units	gpm (gallons/min.)	cfm (cubic ft./min.)
1	3.0	0.041	—	—	—
2	5.0	0.068	—	—	—
3	6.5	0.869	—	—	—
4	8.0	1.069	—	—	—
5	9.4	1.257	5	15.0	2.005
6	10.7	1.430	6	17.4	2.326
7	11.8	1.577	7	19.8	2.646
8	12.8	1.711	8	22.2	2.968
9	13.7	1.831	9	24.6	3.289
10	14.6	1.962	10	27.0	3.609
11	15.4	2.059	11	27.8	3.716
12	16.0	2.139	12	28.6	3.823
13	16.5	2.206	13	29.4	3.930
14	17.0	2.273	14	30.2	4.037
15	17.5	2.339	15	31.0	4.144
16	18.0	2.906	16	31.8	4.241
17	18.4	2.460	17	32.6	4.358
18	18.8	2.513	18	33.4	4.465
19	19.2	2.568	19	34.2	4.572
20	19.6	2.620	20	35.0	4.679
25	21.5	2.874	25	38.0	5.080
30	23.3	3.115	30	42.0	5.614
35	24.9	3.329	35	44.0	5.882
40	26.3	3.516	40	46.0	6.149
45	27.7	3.703	45	48.0	6.417
50	29.1	3.890	50	50.0	6.684
60	32.0	4.278	60	54.0	7.219
70	35.0	4.679	70	58.0	7.753
80	38.0	5.080	80	61.2	8.181
90	41.0	5.481	90	64.3	8.596
100	43.5	5.815	100	67.5	9.023
120	48.0	6.417	120	73.0	9.759
140	52.5	7.018	140	77.0	10.293
160	57.0	7.620	160	81.0	10.828
180	61.0	8.154	180	85.5	11.430
200	65.0	8.689	200	90.0	12.031
225	70.0	9.358	225	95.5	12.766
250	75.0	10.026	250	101.0	13.502
275	80.0	10.694	275	104.5	13.970
300	85.0	11.363	300	108.0	14.437
400	105.0	14.036	400	127.0	16.977
500	124.0	16.576	500	143.0	19.116
750	170.0	22.726	750	177.0	23.661
1000	208.0	27.805	1000	208.0	27.805
1250	239.0	31.950	1250	239.0	31.950
1500	269.0	35.960	1500	269.0	35.960
1750	297.0	39.703	1750	297.0	39.703
2000	325.0	43.446	2000	325.0	43.446
2500	380.0	50.798	2500	380.0	50.798
3000	433.0	57.883	3000	433.0	57.883
4000	535.0	70.182	4000	525.0	70.182
5000	593.0	79.272	5000	593.0	79.272

Source: International Code Council, Inc. See Figure and Table Credits, page 6.18.

Showers and baths have become high-end consumer products. People can now have shower towers, spas, and whirlpool baths installed in their homes (see *Figure 5*). These fixtures demand higher flow rates than normal showers and baths. Consult the local code, because it will specify allowable pressures and flow rates. Codes also specify the proper connection to the supply and recirculation systems. Ensure that the supply system can handle the demand. The product's specifications and installation information will provide the required data.

Next, determine whether the water demand will be intermittent or continuous. Outlets such as lawn faucets, sprinkler systems, and irrigation systems are common continuous-demand items. When they are used, the water flow is constant. The system must supply makeup water. Notice that the fixtures listed above are used only at certain times of the year. Lawn sprinklers and irrigation systems are usually used from the spring to the early fall. Other continuous-demand systems are needed only under special conditions. Fire suppression sprinklers are an example. Be sure to size fire sprinklers so that they can provide enough water in an emergency. Check the product's instructions.

Sinks, lavatories, water closets, and similar devices are used for about five minutes or less at a time. They are used only intermittently. Consult the local code for requirements in your area.

> **WARNING!**
>
> Always use water supply piping that is made from materials specified in the local code. All water supply piping must be resistant to corrosion. The amount of lead allowed in piping materials is strictly limited. Failure to use approved materials could contaminate the potable water supply. This could cause illness and even death.

Combine the amount of flow from continuous- and intermittent-demand fixtures and outlets. Factor this information into the earlier rate of flow estimate.

Calculate the system's maximum probable flow. This number is an estimate of peak water demand. Each fixture has flow and demand pressure values assigned to it. You will learn how to calculate flow and demand later in *Plumbing Level Three*, in the module *Water Pressure Booster and Recirculation Systems*. Consult with local plumbing and code experts to learn about the values used in your area.

Finally, determine the **developed length** of each branch line. The developed length is the total length of piping from the supply to a single fixture. This includes pipes, elbows, valves, tees, and water heaters. Refer back to *Table 1* to determine the equivalent lengths for fittings.

308F05.EPS

Figure 5 ◆ A luxury shower tower installation.

Install valves, regulators, and strainers so that they are accessible because you may need to repair or replace these items periodically. Valves and strainers should be installed so that they can be repaired without removing them.

4.0.0 ◆ SIZING WATER SUPPLY PIPING

Once you have estimated the system's requirements and demands, you can begin to size the system. There is more than one right way to size a water supply system. The method discussed in this section is based on the 2000 International Plumbing Code®. Plumbers in your area may use different engineering practices. Take the time to learn how systems are sized in your area. Whatever method you use, ensure that you follow the highest professional standards. The result will be a water supply system that provides the right amount of water at the right amount of pressure.

Begin with an isometric sketch of the entire system (see *Figure 6*). Then, determine the minimum acceptable pressure to be provided to the highest fixture in the system. If adequate pressure is provided to this fixture, you can assume that there will be enough pressure in the rest of the system. Your local code will specify the minimum pressure for flush valves and flush tanks. Note that the system's pressure requirements must not fall below the minimum pressure or exceed the maximum pressure provided by the water supply.

 DID YOU KNOW?

History of Pipes and Pipe Sizing in the United States

In the 18th and 19th centuries, pipe sizing was an art, not a science. Early plumbers used hollowed-out logs. They were limited to the thickness of the logs they could find, generally 9- to 10-inch-wide elm and hemlock trees. Wooden pipes were not ideal because they often sagged, and insects bred in the stagnant puddles that formed at the low points. Plus, log pipes gave water an unappetizing "woody" flavor.

Vents were among the first pipes to be sized. Until the late 19th century, vents were designed too small and frequently became clogged with ice or debris. In 1874, an unknown plumber invented a vent that balanced system pressure with the outside air pressure. The system used ½-inch pipe, which was wider than pipes used previously. The plumber also extended the pipe outside the building. It worked for a while, but then plugged up. Through constant experimentation, plumbers eventually discovered the proper size of pipe to use in vents. Plumbers were able to apply this knowledge to sizing supply and drain pipes.

Today's standards for pipe sizing have come a long way since the early days of trial and error.

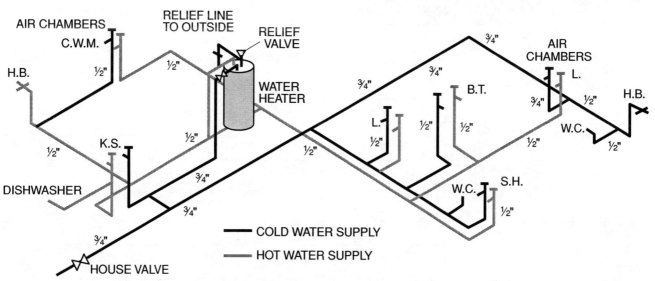

Figure 6 ◆ Water supply piping system isometric drawing.

306F06.EPS

Begin by creating a table, such as the one shown in *Table 5*, to calculate the correct pipe sizes for the system. Divide the isometric drawing of the system (refer to *Figure 6*) into sections. The sections should occur where there are branches or changes in elevation. Add each section to the table. Depending on the system, consider calculating the hot and cold water requirements separately. Enter the flow in gpm for each section.

Find the minimum pressure available from the supply. Then calculate the highest pressure required in the system. Using your local code, cal-culate the pressure drop in the system due to the various fittings. Pressure drop is a loss of water pressure caused by:

- Water meters
- Water main taps
- Water filters and softeners
- Backflow preventers
- Pressure regulators
- Valves and fittings (refer to *Table 1*)
- Pipe friction (see *Figure 7*)

Table 5 Sample Table for Calculating Pipe Sizes

Col	1		2	3	4	5	6	7	8	9	10
Line	**Description**		psi or WSFU	gpm through section	Length of section (ft.)	Trial pipe size (in.)	Equivalent length of fittings and valves (ft.)	Total equivalent length, col. 4 + col. 6 (100 ft.)	Friction loss per 100 feet of trial size pipe (psi)	Friction loss in equivalent length, col. 8 × col. 7	Excess pressure over friction loss (psi)
A	Service and cold water distribution piping[a]	Minimum pressure at main	(psi)								
B		Highest pressure required at fixture	"								
C		Meter loss, 2" meter	"								
D		Tap in main loss, 2" tap	"								
E		Static head loss	"								
F		Backflow preventer loss	"								
G		Filter loss	"								
H		Other loss	"								
I		Total losses and requirements									
J		Pressure available over pipe friction	"								
	Pipe section (from diagram)— Cold water distribution piping		(WSFUs)								
	Total pipe friction losses (cold)										
	Difference (line J – line K)										
	Pipe section (from diagram)— Hot water distribution piping		(WSFUs)								
	Total pipe friction losses (hot)										
	Difference (line J – line K)										

[a] To be considered as pressure gain for fixtures below the main. To consider separately, omit from line I and add to line J.

Source: International Code Council, Inc. See Figure and Table Credits, page 6.18.

PLUMBING LEVEL THREE — TRAINEE MODULE 02306

Your local code will provide tables and graphs to help you calculate pressure drop.

Calculate the *static head loss*, which is the difference in elevation between the supply line and the system's highest fixture. Enter all this information into the table on the appropriate lines and add them together. The result is the total overall losses and requirements.

Subtract the total losses from the minimum pressure available. This is the system pressure that is available to overcome friction loss. You can use this information to select the appropriate pipe size for each section of the system. Note that in some systems, the main is above the highest fixture. In this case, the main pressure must be added to the system, instead of subtracted from it.

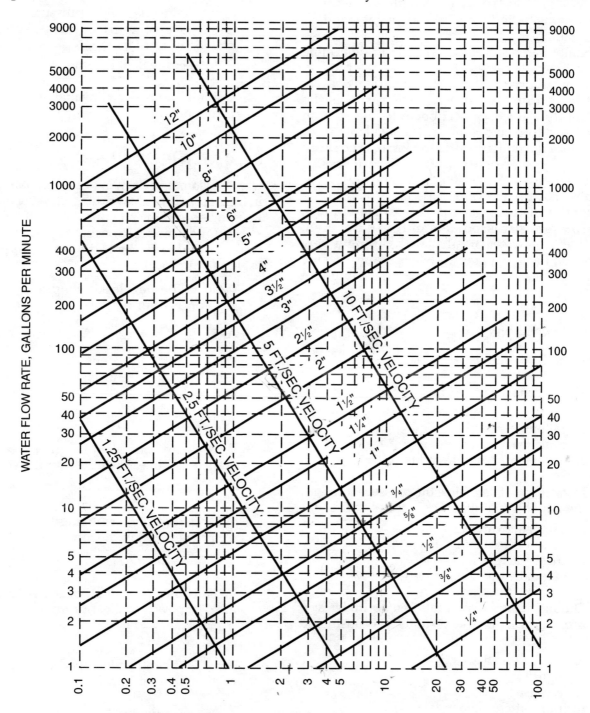

PRESSURE DROP PER 100 FEET OF TUBE, POUNDS PER SQUARE INCH
Note: Fluid velocities in excess of 5 to 8 feet/second are not usually recommended

306F07.EPS

Source: International Code Council, Inc. See Figure and Table Credits, page 6.18.

Figure 7 ◆ Sample table for calculating friction loss in smooth pipe.

Add the lengths of all the sections and select a trial pipe size using the following equation:

$$psi = (\text{available pressure}) \times 100 / (\text{total pipe length})$$

Compare this number to the table on friction loss in fittings (refer to *Table 1*). The table indicates the equivalent lengths for trial pipe size of fittings and valves. Add the section lengths and equivalent lengths to find the total equivalent length of the system.

Finally, refer to your local code to determine the friction loss per 100 feet of pipe (refer to *Figure 7*). Multiply this number by the total equivalent length of each section. The result is the friction loss for each section. Add the friction losses together, and subtract the available pressure from the total pipe friction losses. The result is the excess pressure after friction losses. It should be a small positive number. If it is, that means that the trial size pipe is correct for the system. If the result is a large positive number, then the pipe size is too large and can be reduced. Perform a new set of calculations using a new table and the above steps. Always ensure that the final pipe sizes are not less than the minimum size specified in the local code.

Review Questions

Sections 3.0.0–4.0.0

1. Flow rate is measured in __b__.
   a. drain fixture units
   — b. gallons per minute
   c. cubic feet per minute
   d. water supply fixture units

2. Install __b__ to equalize the pressure throughout a high-rise building's supply system.
   a. balancing valves
   — b. pressure-reducing valves
   c. check valves
   d. manually operated valves

3. Outlets such as lawn faucets, sprinkler systems, and irrigation systems are considered __a__ items.
   — a. continuous-demand
   b. high-flow
   c. low-use
   d. intermittent-demand

4. The difference in elevation between the supply line and the system's highest fixture is called the __d__.
   a. equivalent length
   b. minimum pressure requirement
   — c. pressure drop
   d. static head loss

5. To determine the system pressure that is available to overcome friction loss, subtract the total losses from the __c__.
   a. total equivalent length
   b. static head loss
   — c. minimum pressure available
   — d. friction loss

5.0.0 ◆ THE WORKSHEET

Use the isometric drawing of a plumbing system in the *Appendix* to answer the following questions. Remember to show all your work.

For Questions 1–6, assume ¾-inch piping. Refer to *Table 1* in the text or to the appropriate table in your local code.

1. What is the developed length of piping, in feet, for the four fixtures?

   A. __15__   B. __15.5__   C. __10__   D. __12__

2. What is the equivalent length, in feet, for a ¾-inch check valve installed at each of the four fixtures?

   A. __3__   B. __3__   C. __3__   D. __3__

3. What is the equivalent length, in feet, for the branch tees in each of the four lines?

   A. __0__   B. __3__   C. __6__   D. __3__

4. What is the equivalent length, in feet, for the tee runs in each of the four lines?

   A. __0__   B. __3__   C. __6__   D. __3__

5. What is the equivalent length, in feet, for the elbows in each of the four lines?

   A. __2__   B. __2__   C. __2__   D. __2__

6. What is the total developed length, in feet, of each of the four lines?

   A. __21__   B. __26.5__   C. ____   D. ____

For Questions 7–10, assume a distance of 100 feet from the supply valve. Assume that total friction loss is 10 psi. Refer to *Figure 7* in the text or to the appropriate table in your local code.

7. Find the required pipe size in inches and the velocity in feet per second for Outlet A if it requires 2 gpm.

   Size: 3/8   Velocity: 4

8. Find the required pipe size in inches and the velocity in feet per second for Outlet B if it requires 4 gpm.

   Size: 5/8   Velocity: 5

9. Find the required pipe size in inches and the velocity in feet per second for Outlet C if it requires 10 gpm.

   Size: 3/4   Velocity: 7

10. Find the required pipe size in inches and the velocity in feet per second for Outlet D if it requires 20 gpm.

    Size: 1   Velocity: 8

Summary

A properly sized water supply system ensures that customers have adequate water at the right pressure when they need it. Plumbers install water supply systems tailored for each building. Plumbers must follow several steps and rules when installing a supply system. These apply to all systems, no matter how large or small.

To install a system correctly, plumbers must know about the physical properties of water. These properties include temperature, density, flow, and friction. Before installing the system, plumbers also must find the water requirements for all fixtures and outlets. Then they can estimate the total length of all pipes and fittings. They also calculate how much water the system can handle.

All of this information needs to be collected before the pipes can be sized. Pipe sizing involves several steps. The first is to identify the WSFUs for each branch in the system. Next, the plumber selects the right size of pipe for the branches. The plumber then calculates friction loss in the whole system. With that information, the plumber can select the correctly sized supply line. By carefully following all of these steps, the plumber will be able to install an efficient water supply system.

Trade Terms Introduced in This Module

Demand: The measure of the water requirement for the entire water supply system.

Density: The amount of a liquid, gas, or solid in a space, measured in pounds per cubic foot.

Developed length: The length of all piping and fittings from the water supply to a fixture.

Equivalent length: The length of pipe required to create the same amount of *friction* as a given fitting.

Flow rate: The rate of water flow in gallons per minute that a fixture uses when operating.

Friction: The resistance that results from objects or substances rubbing against one another.

Friction loss: The partial loss of system pressure due to *friction*. Friction loss is also called *pressure loss*.

Laminar flow: The parallel flow pattern of a liquid that is flowing slowly. Also called *streamline flow* or *viscous flow*.

Pressure drop: In a water supply system, the difference between the pressure at the inlet and the pressure at the farthest outlet.

Transient flow: The erratic flow pattern that occurs when a liquid's flow changes from a *laminar flow* to a *turbulent flow* pattern.

Turbulent flow: The random flow pattern of fast-moving water or water moving along a rough surface.

Viscosity: The measure of a liquid's resistance to flow.

Water supply fixture unit: The measure of a fixture's load. The load depends on water quantity and temperature and on fixture type.

Worksheet Illustration

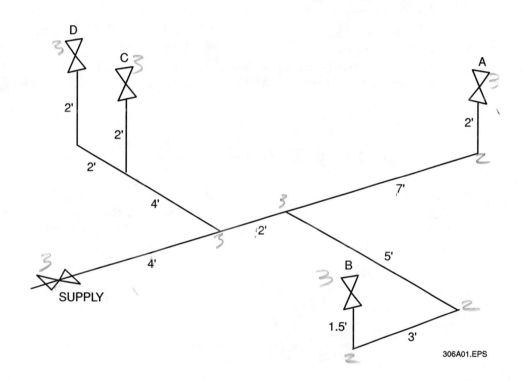

306A01.EPS

Additional Resources

This module is intended to present thorough resources for task training. The following reference works are suggested for further study. These are optional materials for continued education rather than for task training.

Code Check Plumbing: A Field Guide to Plumbing, 2000. Redwood Kardon, Michael Casey, and Douglas Hansen. Newtown, CT: Taunton Press.

Plumbers and Pipefitters Handbook, 1996. William J. Hornung. Englewood Cliffs, NJ: Prentice Hall College Division.

Standard Plumbing Engineering Design, Second Edition, 1982. Louis S. Neilsen. New York, NY: McGraw-Hill.

Figure and Table Credits

Tables 1–5 and Figure 7 (306F07): Written permission to reproduce this material was sought from, and granted by, the copyright holder, **International Code Council, Inc.,** 5203 Leesburg Pike, Suite 600, Falls Church, VA 22041.

Jacuzzi Company   306F05

References

2000 International Plumbing Code, 2000. Falls Church, VA: International Code Council.

Code Check Plumbing: A Field Guide to Plumbing, 2000. Redwood Kardon, Michael Casey, and Douglas Hansen. Newtown, CT: Taunton Press.

NCCER CRAFT TRAINING USER UPDATES

The NCCER makes every effort to keep these textbooks up-to-date and free of technical errors. We appreciate your help in this process. If you have an idea for improving this textbook, or if you find an error, a typographical mistake, or an inaccuracy in the NCCER's Craft Training textbooks, please write us, using this form or a photocopy. Be sure to include the exact module number, page number, a detailed description, and the correction, if applicable. Your input will be brought to the attention of the Technical Review Committee. Thank you for your assistance.

Instructors – If you found that additional materials were necessary in order to teach this module effectively, please let us know so that we may include them in the Equipment and Materials list in the Instructor's Guide.

Write: Curriculum Revision and Development Department
National Center for Construction Education and Research
P.O. Box 141104, Gainesville, FL 32614-1104

Fax: 352-334-0932

E-mail: curriculum@nccer.org

Craft _____ Module Name _____

Copyright Date _____ Module Number _____ Page Number(s) _____

Description _____

_____

_____

_____

(Optional) Correction _____

_____

_____

(Optional) Your Name and Address _____

_____

_____

_____

Backflow Preventers

COURSE MAP

This course map shows all of the modules in the third level of the Plumbing curriculum. The suggested training order begins at the bottom and proceeds up. Skill levels increase as you advance on the course map. The local Training Program Sponsor may adjust the training order.

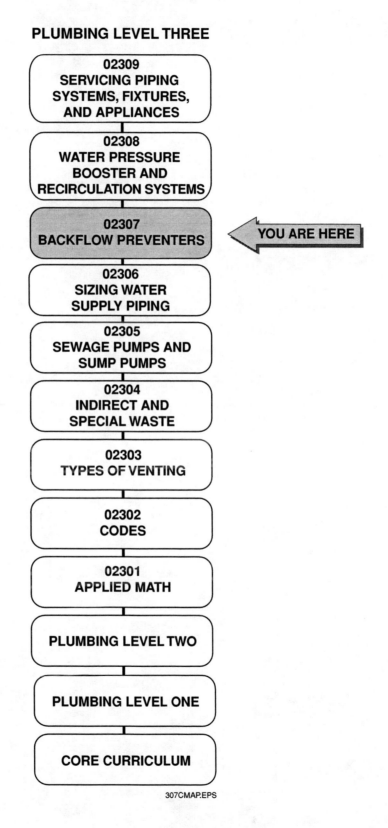

PLUMBING LEVEL THREE

02309
SERVICING PIPING
SYSTEMS, FIXTURES,
AND APPLIANCES

02308
WATER PRESSURE
BOOSTER AND
RECIRCULATION SYSTEMS

02307
BACKFLOW PREVENTERS

YOU ARE HERE

02306
SIZING WATER
SUPPLY PIPING

02305
SEWAGE PUMPS AND
SUMP PUMPS

02304
INDIRECT AND
SPECIAL WASTE

02303
TYPES OF VENTING

02302
CODES

02301
APPLIED MATH

PLUMBING LEVEL TWO

PLUMBING LEVEL ONE

CORE CURRICULUM

307CMAP.EPS

Figures

Table

Backflow Preventers

Objectives

When you have completed this module, you will be able to do the following tasks in accordance with local codes:

1. Explain the principle of backflow due to back siphonage or back pressure.
2. Explain the hazards of backflow and demonstrate the importance of backflow preventers.
3. Identify and explain the applications of the six basic backflow prevention devices.
4. Install common types of backflow preventers.

Prerequisites

Before you begin this module, it is recommended that you successfully complete the following: Core Curriculum; Plumbing Level One; Plumbing Level Two; Plumbing Level Three, Modules 02301 through 02306.

Required Trainee Materials

1. Appropriate personal protective equipment
2. Pencil and paper
3. Copy of your local code

1.0.0 ◆ INTRODUCTION

Plumbing systems deliver fresh water and take away wastewater. Normally, these two water sources are separated. However, sometimes problems or malfunctions can force water to flow backward through a system. When this happens, wastewater and other liquids can be siphoned into the fresh water supply. This can cause contamination, sickness, and even death. Codes require plumbing systems to protect against reverse flow.

Protection can be in the form of a gap or a barrier between the two sources. Check valves can be effective barriers and can be arranged to provide increasing protection against contamination. In this module, you will learn how to install devices that protect potable water from contamination.

2.0.0 ◆ BACKFLOW AND CROSS CONNECTIONS

The reverse flow of nonpotable liquids into the potable water supply is called **backflow.** Backflow can contaminate a fixture, a building, or a community. There are two types of backflow: back pressure and **back siphonage.** Back pressure is a higher than normal downstream pressure in the potable water system. Back siphonage is a lower than normal upstream pressure.

Backflow cannot happen unless there is a direct link between the potable water supply and another source. This condition is called a **cross connection.** Cross connections are not hazards all by themselves (see *Figure 1*). In fact, sometimes they are required. Cross connections are sometimes hard for the public to spot. Many people might not know that a hose left in a basin of wastewater could cause a major health hazard by creating a backflow.

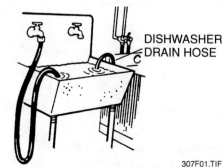

DISHWASHER DRAIN HOSE

307F01.TIF

Figure 1 ◆ Simple cross connection.

When something creates a backflow, a cross connection becomes dangerous. Several things can cause backflow, which, in turn, can force wastes through a cross connection:

- Cuts or breaks in the water main
- Failure of a pump
- Injection of air into the system
- Accidental connection to a high-pressure source

Plumbers use backflow preventers to protect against cross connections. They provide a gap or a barrier to keep backflow from entering the water supply. Many local codes now require backflow preventers for all fixtures in a new structure. Tank trucks filled with sewage, pesticides, or other dangerous substances must also use backflow preventers. Plumbers are responsible for installing and testing backflow preventers in plumbing systems.

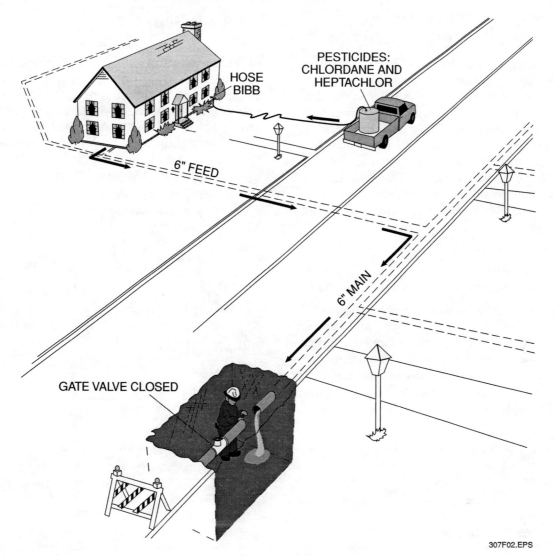

Figure 2 ◆ How dangerous back siphonage can occur.

Review Questions

Sections 1.0.0–2.0.0

1. Backflow is the _____ of nonpotable liquids into the potable water supply.
   a. drainage
   b. reverse flow
   c. pumping
   d. draw off

2. Back pressure is caused by _____ downstream pressure in the potable water system.
   a. higher than normal
   b. equal
   c. lower than normal
   d. positive

3. Back siphonage is caused by _____ upstream pressure in the potable water system.
   a. unequal
   b. excessive
   c. higher than normal
   d. lower than normal

4. One common cause of backflow is _____.
   a. failure of check valves
   b. malfunctioning pumps
   c. clogged drains
   d. improperly located vents

5. Backflow preventers consist of a gap or _____ that prevents backflow in case of a cross connection.
   a. relief valve
   b. barrier
   c. trap
   d. interceptor

3.0.0 ◆ TYPES OF BACKFLOW PREVENTERS

Backflow preventers help keep drinking water safe. They are an important part of a plumbing system because they help maintain public health. There are six types of basic backflow preventers, each designed to work under different conditions. Some protect against both back pressure and back siphonage, while others can handle only one type of backflow. Before installing a backflow preventer, always ensure that it is appropriate for the type of installation. Consider several factors:

- Risk of cross connection
- Type of backflow
- Health risk posed by pollutants or contaminants

Preventers must be installed correctly. Select the proper preventer for the application, and review the manufacturer's instructions before installing a preventer. Never attempt to install a backflow preventer in a system for which it is not designed because it may fail, resulting in backflow. Refer to your local code for specific guidance.

3.1.0 Air Gaps

An air gap is a physical separation between a potable water supply line and the flood-level rim of a fixture (see *Figure 3*). It is the simplest form of backflow preventer. Ensure that the supply pipe terminates above the flood-level rim at a distance that is at least twice the diameter of the pipe. The air gap should never be less than 1 inch. Some codes have specific requirements for fixtures with openings that are less than 1 inch (see *Table 1*).

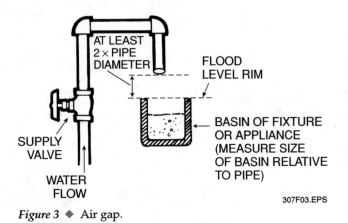

Figure 3 ◆ Air gap.

Table 1 Minimum Air Gaps From a Model Code

Fixture	Minimum Air Gap	
	Away from wall (inches)	Close to wall (inches)
Lavatories and other fixtures with an effective opening not greater than ½ inch in diameter	1	1½
Sinks, laundry trays, gooseneck back faucets, and other fixtures with effective openings not greater than ¾ inch in diameter	1½	2½
Over-rim bath fillers and other fixtures with effective openings not greater than 1 inch in diameter	2	3
Drinking water fountains, single orifice not greater than ⁷⁄₁₆ inch in diameter or multiple orifices with a total area of the same area	1	1½
Effective openings greater than 1 inch	Two times the diameter of the effective opening	Three times the diameter of the effective opening

ON THE ·LEVEL·

Backflow Preventers in Restaurants

Many fixtures and appliances in restaurants depend on clean, fresh water. Steamers, dishwashers, coffee machines, icemakers, and soda dispensers all connect to the fresh water supply. They all require backflow preventers. Public health inspectors regularly test these systems to ensure that they are working properly. A restaurant in violation of the local health code can be shut down. An accidental cross connection, therefore, could not only cause sickness and death but could also damage a restaurant's reputation and threaten its economic survival.

All faucets must incorporate an air gap. Air gaps can be used to prevent back pressure, back siphonage, or both. Never submerge hoses or other devices connected to the potable water supply in a basin that may contain contaminated liquid. Even seemingly innocent activities such as filling up a wading pool with a garden hose could be a problem if, elsewhere, workers digging with a backhoe accidentally cut through the water main.

3.2.0 Atmospheric Vacuum Breakers

If it is not possible to install an air gap, consider using an **atmospheric vacuum breaker** (see *Figure 4*). Atmospheric vacuum breakers (AVBs) use a silicon disc float to control water flow. In normal operation, the float rises to allow potable water to enter. If back siphonage causes low pressure, air enters the breaker from vents and forces the disc into its seat. This shuts down the flow of fresh water (see *Figure 5*).

307F04.EPS

Figure 4 ◆ Cutaway of an atmospheric vacuum breaker.

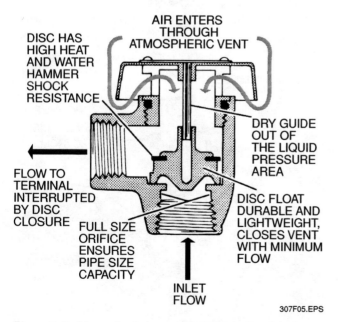

307F05.EPS

Figure 5 ◆ Atmospheric vacuum breaker in the closed position.

Use AVBs for nonpotable water systems that are not in constant use, such as the following:

- Commercial dishwashers
- Sprinkler systems
- Outdoor faucets

AVBs can be used where a pipe or hose terminates below the flood-level rim of a basin (see *Figure 6*).

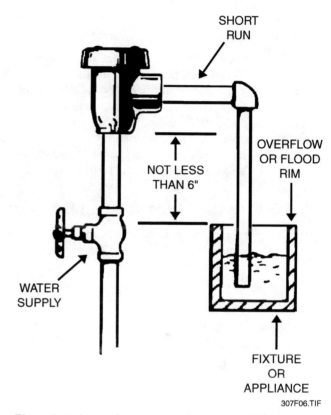

307F06.TIF

Figure 6 ◆ Atmospheric vacuum breaker installation.

AVBs are designed to operate at normal air pressure. Do not use them where there is a risk of back pressure backflow. Check the manufacturer's specifications for the proper operating temperatures and pressures. If conditions exceed these limits, consult the manufacturer first.

Do not install an AVB where it will be exposed to higher than normal air pressure. Install an AVB after the last control valve, and ensure that it is at least 6 inches above the fixture's flood-level rim. Install the breaker so that the fresh water intake is at the bottom, and ensure that the water supply is flowing in the same direction as the arrow on the breaker.

Hose connection vacuum breakers (see *Figure 7*) are designed for use where there is a hose connection. Some codes require them when there is a risk of back siphonage in hoses attached to any of the following:

- Sill cocks (hose bibbs)
- Laundry tubs
- Service sinks
- Photograph developing sinks
- Dairy barns
- Wash racks

Hose connection vacuum breakers work the same way and under the same conditions as AVBs. Note that hose connections do not allow water to drain from an outdoor faucet that is not protected from freezing. If there is a danger of freezing, install a non-freeze faucet with a built-in vacuum breaker. This will allow the faucet to drain. Both types of AVBs can be equipped with strainers.

3.3.0 Pressure-Type Vacuum Breakers

Install **pressure-type vacuum breakers** (see *Figure 8*) to prevent back siphonage, back pressure, or both. They should be used in water lines that supply the following:

- Cooling towers
- Swimming pools
- Heat exchangers
- Degreasers
- Lawn sprinkler systems with downstream zone valves

A pressure-type vacuum breaker (PVB) has a spring-loaded check valve and a spring-loaded air inlet valve. The PVB operates when line pressure drops to atmospheric pressure or less. The float valve opens the vent to the outside air (see *Figure 9*). The check valve closes to prevent back siphonage. PVBs are suitable for continuous pressure applications. PVBs are equipped with

test cocks, which are valves that allow testing of the individual pressure zones within the device. They can also be equipped with strainers if required.

3.4.0 Dual-Check Valve Backflow Preventers

Homes are full of potential sources of contamination. The most common include the following:

- Hose-attached garden spray bottles
- Lawn sprinkler systems
- Bathtub whirlpool adapters
- Hot tubs
- Water closet bowl deodorizers
- Wells or back-up water systems
- Photography darkrooms
- Misapplied pest extermination chemicals

Figure 8 ◆ Pressure-type vacuum breaker.

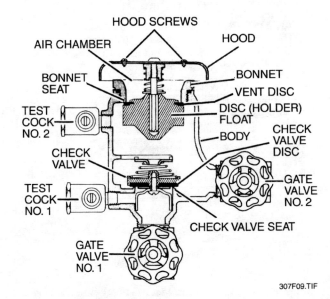

Figure 9 ◆ Cutaway illustration of pressure-type vacuum breaker.

Figure 7 ◆ Hose connection vacuum breaker.

To keep residential backflow from reaching the water main, use a **dual-check valve backflow preventer** (DC) (see *Figure 10*). These preventers feature two spring-loaded check valves. They work similarly to **double-check valve assemblies** (DCVs).

DCs protect against back pressure and back siphonage. Because DCs are smaller, they are commonly used in residential installations. These preventers tend to be less reliable than DCVs. Unlike DCVs, they are normally not fitted with shutoff valves or test cocks. This makes it difficult for plumbers to remove, test, or replace them. Check the local code before installing DCs. Install DCs on the customer side of a residential water meter. They can be installed either horizontally or vertically, depending on the available space and the existing piping arrangements.

3.5.0 Double-Check Valve Assemblies

Where heavy-duty protection is needed, install a double-check valve assembly to protect against back siphonage, back pressure, or both (see *Figure 11*). This type of backflow preventer has two spring-loaded check valves and includes a test cock. They are designed for installation where both back pressure and back siphonage are threats. During normal operation, both check valves are open; however, backflow causes the valves to seal tightly.

DCVs are available in a wide range of sizes from ¾ inch up to 10 inches. DCVs have a gate or ball shutoff valve at each end. If required, DCV inlets can be equipped with strainers.

3.6.0 Reduced-Pressure Zone Principle Backflow Preventers

Reduced-pressure zone principle backflow preventers (RPZs) protect the water supply from contamination by dangerous liquids (see *Figure 12*). RPZs are the most sophisticated type of backflow preventer. They feature two spring-loaded check valves plus a hydraulic, spring-loaded pressure differential relief valve. The relief valve is located underneath the first check valve. The device creates a low pressure between the two check valves. If the pressure differential between the inlet and the space between the check valves drops below a specified level, the relief valve opens. This allows water to drain out of the space between the check valves (see *Figure 13*). The relief valve will also operate if one or both check valves fail. This offers added protection against backflow. RPZs also have shutoff valves at each end.

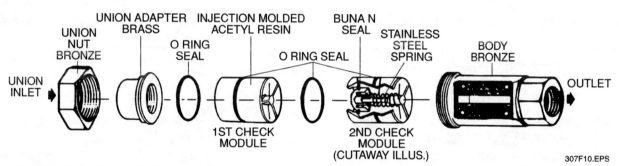

Figure 10 ◆ Dual-check valve backflow preventer.

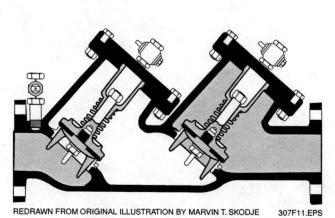

REDRAWN FROM ORIGINAL ILLUSTRATION BY MARVIN T. SKODJE   307F11.EPS

Figure 11 ◆ Double-check valve assembly.

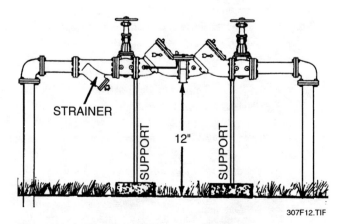

Figure 12 ◆ Installation of a reduced-pressure zone principle backflow preventer.

REDUCED-PRESSURE PRINCIPLE OPERATION
STATIC PRESSURE

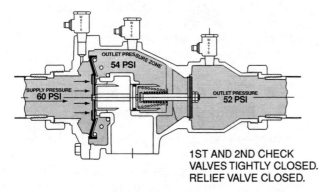

1ST AND 2ND CHECK
VALVES TIGHTLY CLOSED.
RELIEF VALVE CLOSED.

REDUCED-PRESSURE PRINCIPLE OPERATION
BACK PRESSURE

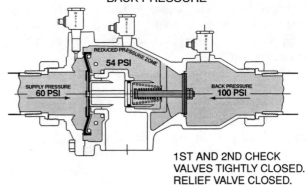

1ST AND 2ND CHECK
VALVES TIGHTLY CLOSED.
RELIEF VALVE CLOSED.

REDUCED-PRESSURE PRINCIPLE OPERATION
FULL FLOW

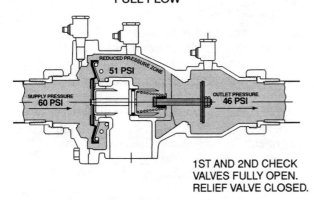

1ST AND 2ND CHECK
VALVES FULLY OPEN.
RELIEF VALVE CLOSED.

REDUCED-PRESSURE PRINCIPLE OPERATION
BACK SIPHONAGE

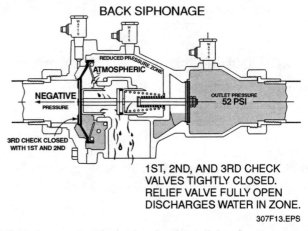

1ST, 2ND, AND 3RD CHECK
VALVES TIGHTLY CLOSED.
RELIEF VALVE FULLY OPEN
DISCHARGES WATER IN ZONE.

307F13.EPS

Figure 13 ◆ Operation of a reduced-pressure zone principle backflow preventer with check valves parallel to flow.

DID YOU KNOW?

E. Coli *and Legionnaire's Disease—Two Dangers Spread by Cross Connection*

Bacteria are safe when they stay where they belong. When they get into the water supply, they can be deadly. *Escherichia coliform,* also known as *E. coli,* is a common type of bacteria. It can be found in animal and human intestines, where it provides natural vitamins, but *E. coli* in animal wastes can be dangerous. It can enter the fresh water supply through untreated runoff and cross connections. *E. coli* turns safe potable water into a disease delivery system and can contaminate meat and other foods. Because plumbers are responsible for installing and maintaining safe plumbing systems, *E. coli* is a real concern for them, too.

Health officials regularly test for *E. coli* outbreaks in the water supply. When one occurs, it can cause chaos and severe illness. Residents have to use bottled waters or tanker trucks have to bring in water for drinking and cooking. Contaminated water must be boiled before people can use it to even brush their teeth. Testing, sampling, and cleanups can take weeks. Sometimes months go by before the health department declares a system is completely clean.

Bacteria cause other health problems, including Legionnaire's disease. The first major reported outbreak of this disease occurred at an American Legion conference in 1976. Scientists named the bacteria Legionella. It reproduces in warm water. Legionella can grow in hot water tanks and some plumbing pipes. People inhale the bacteria through contaminated water mist from large air conditioning installations, whirlpool spas, and showers. It is not contagious, but between 5 and 30 percent of people who contract Legionnaire's disease die from it. Antibiotics do not kill Legionella, but they can keep the bacteria from multiplying, which helps a patient's immune system to eventually kill them.

Plumbers can help block the spread of disease caused by waterborne bacteria. Design plumbing systems so that cross connections can't happen. Install interceptors and pumps to keep out harmful waste and to prevent water from stagnating. Plumbers play a vital role in maintaining public health. Doctors can treat disease, but plumbers can help prevent it.

Use an RPZ if a direct connection is subject to back siphonage, back pressure, or both. A wide variety of industrial and agricultural operations require RPZs, including:

- Car washes
- Poultry farms
- Dairies
- Medical and dental facilities
- Wastewater treatment plants
- Manufacturing plants

Some codes require RPZs on additional types of installations. RPZs are available for all operating temperatures and pressures. Refer to your local code to see where RPZs are required. Many RPZs come with union or flanged connections, which allow fast and easy removal for servicing. Install RPZs at least 12 inches above grade, and provide support for the RPZ (see *Figure 14*). Otherwise, sagging could damage the RPZ. Select a safe location that will help prevent damage and vandalism of the RPZ. Install a **manufactured air gap** on all RPZ installations (see *Figure 15*). It acts as a drain receptor to prevent siphonage.

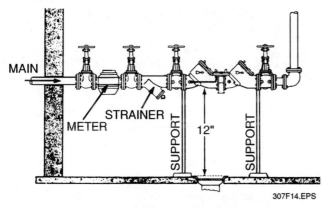

Figure 14 ◆ Reduced-pressure zone principle backflow preventers installed with supports.

Figure 15 ◆ Manufactured air gap.

Some codes permit two RPZs to be installed in a parallel fashion (see *Figure 16*). This allows the system to continue working if an RPZ malfunctions or is being serviced. If two RPZs are used in parallel, a smaller size can be used than if only one RPZ were installed. Locate a strainer in the pipe before both backflow preventers. If there is sediment in the system, it may clog a single strainer. In that case, install separate strainers in advance of each RPZ. Consult with the project engineer and the local plumbing code official before installing parallel backflow preventers.

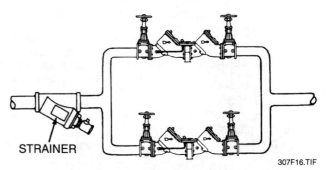

Figure 16 ◆ Reduced-pressure zone principle backflow preventers installed in parallel.

Constant water dripping from an RPZ indicates that the valve disc is not seating properly. Flush the unit, and the problem should correct itself. If it does not, refer to manufacturer's specifications for proper maintenance procedures.

4.0.0 ◆ SPECIALTY BACKFLOW PREVENTERS

Often, low-flow and small supply lines require backflow preventers. Use an **intermediate atmospheric vent vacuum breaker** for this kind of line (see *Figure 17*). The intermediate atmospheric vent is a form of PVB that also protects against back pressure. This type of breaker is therefore suitable for use on the following installations:

- Most laboratory equipment
- Processing tanks
- Sterilizers
- Residential boiler feeds

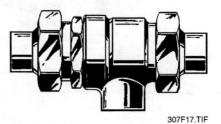

Figure 17 ◆ Intermediate atmospheric vent vacuum breaker.

Laboratories and hospitals use chemicals that can cause illness or death. Backflow preventers help protect the fresh water supply. Laboratory faucets with hose attachments require **in-line vacuum breakers** (see *Figure 18*). In-line vacuum breakers work the same way as AVBs. A disc allows fresh water to flow to the faucet. Negative pressure causes the disc to close the water inlet. In-line vacuum breakers protect against back siphonage. Install them on new and existing faucets. Changes to the plumbing are not required.

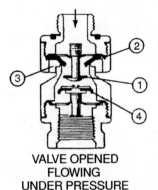

With flow through valve, primary check (1) opens away from diaphragm seal (2). Atmospheric port (3) remains closed by deflection of diaphragm seal. Secondary check (4) opens away from downstream seat, permitting flow of water through valve.

VALVE OPENED
FLOWING
UNDER PRESSURE

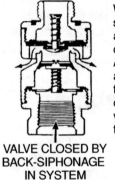

With a back-siphonage condition, secondary check seals tightly against downstream seat. Primary check seals against diaphragm. Atmospheric port is open, permitting air to enter air break chamber. In the event of fouling of downstream check valve, leakage would be vented to the atmosphere through the vent port.

VALVE CLOSED BY
BACK-SIPHONAGE
IN SYSTEM

307F18.EPS

Figure 18 ◆ In-line vacuum breaker.

Review Questions

Sections 3.0.0–4.0.0

1. There are _____ types of basic backflow preventers.
   a. four
   b. five
   c. six
   d. seven

2. An air gap should never be less than _____ inch(es).
   a. ¼
   b. ½
   c. 1
   d. 2

3. Protect the house side of a residential water meter by installing a _____.
   a. hose connection vacuum breaker
   b. dual-check valve backflow preventer
   c. double-check valve assemblies
   d. reduced-pressure zone principle backflow preventer

4. _____ only protect against back siphonage.
   a. Atmospheric vacuum breakers
   b. Air gaps
   c. Double-check valve assemblies
   d. Reduced-pressure zone principle backflow preventers

5. Use an intermediate atmospheric vent pressure-type vacuum breaker on _____ supply lines.
   a. high-pressure (25 to 40 psi)
   b. large (¾- and 1-inch pipe)
   c. contaminant (medium to high risk)
   d. small (½- and ¾-inch pipe)

ON THE

· LEVEL ·

Selecting Backflow Preventers

When determining which type of backflow preventer to install, you must always consider its application. Because backflow preventers help to maintain public health, liability risks affect how potential hazards are defined. The type of installation varies depending on whether these contaminants are defined as low, medium, or high hazard. Therefore, always check your local code before installing a backflow preventer. Make sure you are using the correct device for the installation's requirements.

5.0.0 ◆ TESTING BACKFLOW PREVENTERS

CAUTION

Under no circumstances should an uncertified plumber attempt to do this work.

Backflow preventers can break down. When they do, they need to be repaired or replaced. Many state codes also require regular testing of preventers. Only certified technicians can service and test backflow preventers. Certified testers are licensed in each state. Refer to this section for reference only. The information is not intended to be a step-by-step testing method. Test equipment and procedures will vary.

Many backflow preventers have test cocks, which are used for testing (see *Figure 19*). Foreign matter blocking the valves can cause a malfunction, so many backflow preventers have strainers to trap dirt and other matter. Strainers can be cleaned without having to disconnect the preventer. If sediment interferes with the operation of a backflow preventer, the system needs to be flushed.

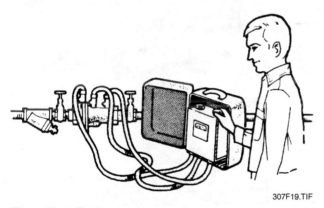

307F19.TIF

Figure 19 ◆ Testing pressure zones.

5.1.0 Testing Atmospheric and Pressure-Type Vacuum Breakers

Certified technicians use sight tube test kits to inspect the air inlets and check valves on vacuum breakers (see *Figure 20*). Sight tubes are usually made of 1-inch diameter clear plastic about 40 inches long. A gauge can also be used to test these breakers. Watch to see if the poppet in the air inlet unseats. If the air inlet does not open, it is sticking and needs to be repaired.

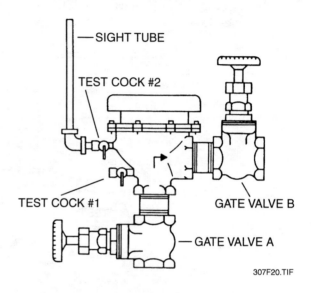

— SIGHT TUBE

TEST COCK #2

TEST COCK #1

GATE VALVE B

— GATE VALVE A

307F20.TIF

Figure 20 ◆ Test equipment installation on a vacuum breaker.

The check valve should be able to withstand a pressure of 1 psi in the direction of flow. The level of water in the sight tube may also drop slightly, but it should not drop below 28 inches above the check valve. If the water in the sight tube continues to drop, the check valve is leaking and will need to be repaired.

ON THE · LEVEL ·

Certified Backflow Preventer Assembly Testers

In many states, plumbers can install backflow preventers. However, preventer testing must be left to certified testers. Codes provide stiff penalties for a noncertified plumber who carries out an inspection. Refer to the local code. Testers must take written and practical exams offered at a site authorized by the local chapter of the American Water Works Association (AWWA). Applicants must have a high school diploma or General Education Development (GED) equivalent, be at least 18 years old, and speak and understand English. In many states, testers do not have to be licensed plumbers. The certification is good for 3 years.

5.2.0 Testing Reduced-Pressure Zone Principle Backflow Preventers

Certified technicians use differential pressure gauge test kits to test the operation of the differential pressure relief valve and both check valves (see *Figure 21*). The differential pressure should be no less than the specified amount (usually 2 psi).

Watch the reading on the differential pressure gauge. The pressure should remain at least 3 psi, or other specified amount, above the relief valve. If it drops below the specified level, the check valve may be leaking and may require repair. The differential pressure should remain above the pressure of the relief valve opening.

5.3.0 Testing Double-Check Valves

Certified technicians test the tightness of DCVs using two test gauges and fittings (see *Figure 22*). The test procedure is basically the same for each check valve. The main difference is the test cocks being used. If a small difference in pressure can be maintained (about 2 psi) and if the gauge shows an increase in the pressure differential, the check valve is in working condition. If not, the check valve may be leaking.

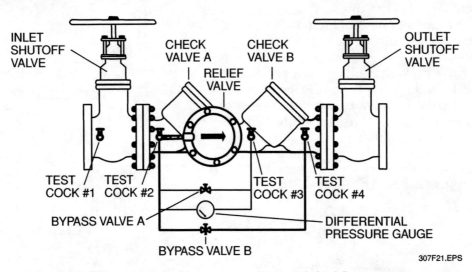

Figure 21 ◆ Test equipment installation on a reduced-pressure zone principle backflow preventer.

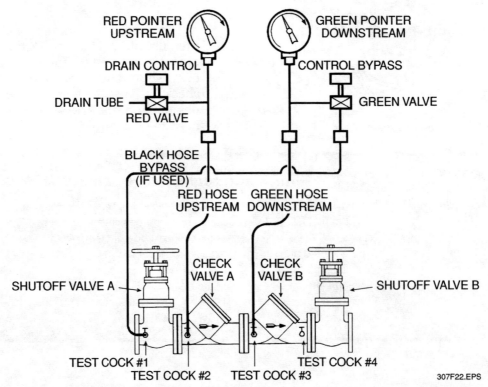

Figure 22 ◆ Test equipment installation on a double-check valve.

Review Questions

Section 5.0.0

1. Certified technicians must be called to service and test backflow preventers. Certified testers are licensed in each _____.
   a. city
   b. county
   c. state
   d. country

2. Many backflow preventers have _____, which are used for testing.
   a. check valves
   b. test cocks
   c. pressure gauges
   d. relief valves

3. In atmospheric and pressure-type backflow preventers, if the water in the sight tube continues to _____, the check valve is leaking.
   a. increase
   b. change
   c. decrease
   d. remain the same

4. In a reduced-pressure zone principle backflow preventer, the pressure in the check valve should remain _____ the pressure in the relief valve.
   a. proportional to
   b. the same as
   c. lower than
   d. higher than

5. In double-check valves, there should be _____ in the pressure differential between the check valves.
   a. a decrease
   b. an increase
   c. a gradual change
   d. no change

Summary

Problems and malfunctions in a plumbing system can force water to flow backward through a system. Backward flow can draw contaminants into a plumbing system through a cross connection. This is called backflow. Back pressure and back siphonage are the two kinds of backflow. Backflow preventers protect the potable water supply from contamination. There are six main types of backflow preventers: air gap, atmospheric vacuum breaker, pressure-type vacuum breaker, double-check valve, dual-check valve, and reduced-pressure zone principle. Each of these is designed for specific applications, and some work on both types of backflow.

Codes require certified technicians to service and test backflow preventers. Do not attempt to service or test a preventer without certification. Regular tests ensure that backflow preventers continue to operate correctly. Always consult with the project engineer and local code officials before selecting and installing a backflow preventer. Public safety depends on the reliable operation of backflow preventers.

Trade Terms Introduced in This Module

Atmospheric vacuum breaker: A *backflow preventer* designed to prevent *back siphonage* by allowing air pressure to force a ceramic disc into its seat and block the flow.

Back siphonage: A form of *backflow* caused by lower than normal upstream pressure in the fresh water line.

Backflow: An undesirable condition that results when nonpotable liquids enter the potable water supply by reverse flow through a *cross connection.*

Cross connection: A direct link between the potable water supply and water of unknown or questionable quality.

Double-check valve assembly (DCV): A *backflow preventer* that prevents *back pressure* and *back siphonage* through the use of two spring-loaded check valves. DCVs are larger than *dual-check valve backflow preventers* and are used for heavy-duty protection. DCVs have test cocks.

Dual-check valve backflow preventer (DC): A *backflow preventer* that uses two spring-loaded check valves to prevent *back siphonage* and *back pressure.* DCs are smaller than *double-check valve assemblies* and are used in residential installations. DCs do not have test cocks.

Hose connection vacuum breaker: A *backflow preventer* designed to be used outdoors on hose bibbs to prevent *back siphonage.*

In-line vacuum breaker: A specialty *backflow preventer* that uses a disc to block flow when subjected to *back siphonage.* It works the same as an *atmospheric vacuum breaker.*

Intermediate atmospheric vent vacuum breaker: A specialty *backflow preventer* used on low-flow and small supply lines to prevent *back siphonage* and *back pressure.* It is a form of *pressure-type vacuum breaker.*

Manufactured air gap: An air gap that can be installed on a *reduced-pressure zone principle backflow preventer* to prevent *back pressure* and *back siphonage.*

Pressure-type vacuum breaker: A *backflow preventer* installed in a water line that uses a spring-loaded check valve and a spring-loaded air inlet valve to prevent *back siphonage.*

Reduced-pressure zone principle backflow preventer: A *backflow preventer* that uses two spring-loaded check valves plus a hydraulic, spring-loaded pressure differential relief valve to prevent *back pressure* and *back siphonage.*

Test cock: A valve in a *backflow preventer* that permits the testing of individual pressure zones.

Additional Resources

This module is intended to present thorough resources for task training. The following reference works are suggested for further study. These are optional materials for continued education rather than for task training.

Backflow Prevention: Theory and Practice, 1990. Robin L. Ritland. Dubuque, IA: Kendall/Hunt Publishing Company.

Manual of Cross-Connection Control, Ninth Edition, 1993. Foundation for Cross-Connection Control and Hydraulic Research. Los Angeles, CA: University of Southern California.

Recommended Practice for Backflow Prevention and Cross-Connection Control, Second Edition, 1990. Denver, CO: American Water Works Association.

Acknowledgments

Eddie Pope, Plumb Works, Atlanta, GA

Watts Regulator Company

References

"50 Cross-Connection Questions, Answers, and Illustrations Relating to Backflow Prevention Products and Protection of Safe Drinking Water Supply," 2001. Publication F-50, Watts Regulator Company.

1997 International Plumbing Code, 1998. Falls Church, VA: International Code Council.

Cross-Connection Control Manual, 1989. Publication 570/9-89-007, United States Environmental Protection Agency, Office of Ground Water and Drinking Water, Washington, DC.

"Questions and Answers About Cross-Connection Control," Florida Department of Environmental Protection website www.mindspring.com/~loben/dep-info.htm, 2001. Florida Department of Environmental Protection.

"Protect Your Water Supply from Agricultural Chemical Backflow," 1993. Extension Bulletin E-2349, Robert H. Wilkinson and Julie Stachecki, Michigan State University Cooperative Extension Service, East Lansing, MI.

Figure Credits

FEBCO   307F20, 307F21, 307F22

U.S. EPA Office of Ground   307F11
Water and Drinking Water

Watts Regulator Company   307F01, 307F02, 307F03,
                                   307F04, 307F05, 307F06,
                                   307F07, 307F08, 307F10,
                                   307F12, 307F14, 307F15,
                                   307F16, 307F17, 307F18,
                                   307F19

NCCER CRAFT TRAINING USER UPDATES

The NCCER makes every effort to keep these textbooks up-to-date and free of technical errors. We appreciate your help in this process. If you have an idea for improving this textbook, or if you find an error, a typographical mistake, or an inaccuracy in the NCCER's Craft Training textbooks, please write us, using this form or a photocopy. Be sure to include the exact module number, page number, a detailed description, and the correction, if applicable. Your input will be brought to the attention of the Technical Review Committee. Thank you for your assistance.

Instructors – If you found that additional materials were necessary in order to teach this module effectively, please let us know so that we may include them in the Equipment and Materials list in the Instructor's Guide.

Write:	Curriculum Revision and Development Department
	National Center for Construction Education and Research
	P.O. Box 141104, Gainesville, FL 32614-1104
Fax:	352-334-0932
E-mail:	curriculum@nccer.org

Craft _____ Module Name _____

Copyright Date _____ Module Number _____ Page Number(s) _____

Description _____

_____

_____

_____

(Optional) Correction _____

_____

_____

(Optional) Your Name and Address _____

_____

_____

Water Pressure Booster and Recirculation Systems

COURSE MAP

This course map shows all of the modules in the third level of the Plumbing curriculum. The suggested training order begins at the bottom and proceeds up. Skill levels increase as you advance on the course map. The local Training Program Sponsor may adjust the training order.

PLUMBING LEVEL THREE

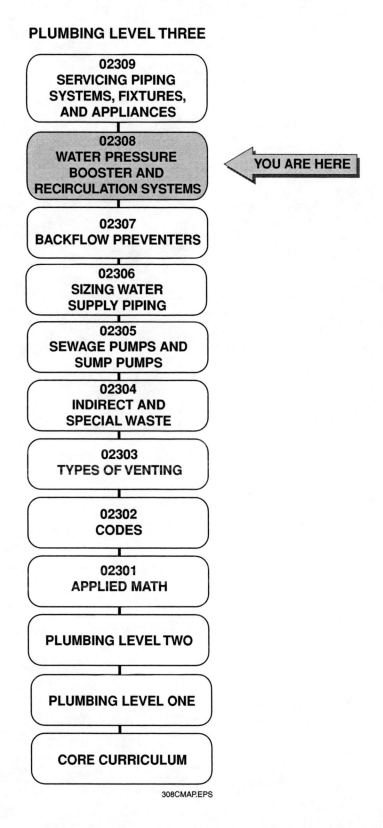

02309
SERVICING PIPING
SYSTEMS, FIXTURES,
AND APPLIANCES

02308
WATER PRESSURE
BOOSTER AND
RECIRCULATION SYSTEMS ← YOU ARE HERE

02307
BACKFLOW PREVENTERS

02306
SIZING WATER
SUPPLY PIPING

02305
SEWAGE PUMPS AND
SUMP PUMPS

02304
INDIRECT AND
SPECIAL WASTE

02303
TYPES OF VENTING

02302
CODES

02301
APPLIED MATH

PLUMBING LEVEL TWO

PLUMBING LEVEL ONE

CORE CURRICULUM

308CMAP.EPS

Figures

Tables

Water Pressure Booster and Recirculation Systems

Objectives

When you have completed this module, you will be able to do the following tasks in accordance with local codes:

1. Explain the complete water pressure booster system and its components.
2. Explain the maintenance and basic troubleshooting processes for water pressure booster systems.
3. Describe the characteristics of the different recirculation systems.
4. Identify the basic components of a recirculation system.
5. Identify the location of various components within a recirculation system.
6. Install a water pressure booster system per engineering plans and specifications.
7. Install the basic components of a recirculation system.
8. Use the local plumbing code to find and cite requirements for recirculation systems.
9. Diagnose basic problems in recirculation systems.

Prerequisites

Before you begin this module, it is recommended that you successfully complete the following: Core Curriculum; Plumbing Level One; Plumbing Level Two; Plumbing Level Three, Modules 02301 through 02307.

Required Trainee Materials

1. Appropriate personal protective equipment
2. Pencil and paper
3. Copy of your local code

1.0.0 ◆ INTRODUCTION

Plumbing allows people to have clean, fresh water on demand. People have come to expect it wherever they live, work, and play. Every day, plumbers are faced with the challenge of designing and installing cold and hot water systems for every type of structure imaginable—single-floor homes, high-rise apartments, office skyscrapers, hotels, shopping centers, and many others. Each of these systems has unique design requirements, and tall buildings present a special challenge. However, each system must perform the same basic functions: it must be able to provide water at the right pressure and temperature.

In this module, you'll learn how to design and install two kinds of water supply systems. One system boosts water pressure in places where gravity limits the flow. The other system provides hot water instantly throughout a building. These systems use pumps, storage tanks, controls, and standard pipes and fittings. You are already familiar with many of the tools and techniques required.

2.0.0 ◆ WATER PRESSURE BOOSTER SYSTEMS

Sometimes, water supply systems need help providing potable water at the right pressure. Many rural houses rely on deep wells for potable water, but the water pressure alone may not be enough to raise the water to the houses. Tall city office buildings can connect to the city water supply, but the city water pressure may not be enough to raise water to the top floor. Plumbers install **water pressure booster systems** to solve problems like these. Water pressure booster systems do exactly what their name suggests. They boost water pressure up to the level required by the plumbing system. Plumbers design booster systems to meet the specific needs of individual buildings.

2.1.0 Types of Water Pressure Booster Systems

There are four main types of water pressure booster systems:

- **Elevated tank systems**
- **Hydropneumatic tank systems**
- **Constant speed systems**
- **Variable capacity systems**

Each system was designed as an improvement over the previous one, and now earlier systems are considered outdated because of new technology and standards. Each of these systems is discussed in more detail below.

2.1.1 Elevated Tank Systems

The elevated tank system is the simplest type of water pressure booster system. It is also the oldest. In an elevated tank system, a water tank is installed on the top floor of a building. A small pump fills the tank. The weight of the water column provides the required pressure to the fixtures below (see *Figure 1*). Elevated tanks are also called *gravity tanks*.

Elevated tank systems are no longer used. Many years ago, they were well suited for smaller buildings not connected to a city service. However, they are not effective in tall buildings or compatible with modern plumbing systems. In addition, the weight of a large water tank on a tall building can cause severe structural problems.

2.1.2 Hydropneumatic Tank Systems

The next major innovation in water pressure booster systems was the hydropneumatic tank system. These systems are used in smaller installations, such as in the floor of a building. Air pressure forces water from a tank to various parts of a building (see *Figure 2*). These systems have several drawbacks. The constant contact between air and water can cause excessive corrosion, which can cause unsanitary conditions in the system. Over time, air diffuses into the water, which causes the tank to become waterlogged and reduces the system pressure. Water in the plumbing system is also at risk from contamination by compressor oil. These problems can plague older systems. However, newer and more sophisticated systems are less likely to have them.

Install the largest hydropneumatic tank possible for the installation's needs. Select a pump that will be able to charge the tank fully. The tank should be able to maintain a minimum of a 10 percent water seal at full air volume. The system's pressure variation should be as large as possible. The system pressure will be highest when the tank is full, and it will be lowest when the tank is nearly empty. Set the pump to activate when the tank reaches a specified low pressure. Consult the tank specifications for this information. The goal is to design a system that operates economically and efficiently. Select a variation that makes sense. Obviously, a large pressure variation at a faucet or a drinking fountain is not very desirable!

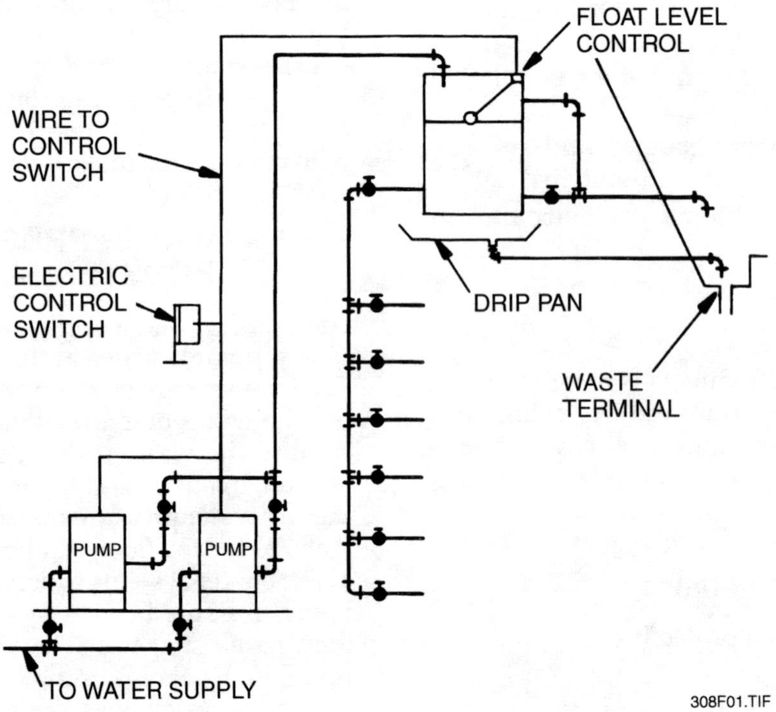

Figure 1 ◆ Schematic of a gravity tank system.

PLUMBING LEVEL THREE — TRAINEE MODULE 02308

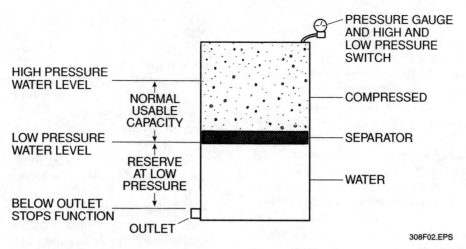

HIGH PRESSURE WATER LEVEL

NORMAL USABLE CAPACITY

LOW PRESSURE WATER LEVEL

RESERVE AT LOW PRESSURE

BELOW OUTLET STOPS FUNCTION

OUTLET

PRESSURE GAUGE AND HIGH AND LOW PRESSURE SWITCH

COMPRESSED

SEPARATOR

WATER

308F02.EPS

Figure 2 ◆ Hydropneumatic tank.

2.1.3 *Constant-Speed Systems*

In a constant-speed system, water is pumped continuously through the plumbing installation (see *Figure 3*). When water is used, the system pumps replacement water from the main. This system does not use air to create the water pressure. This eliminates the risk of corrosion, which can affect hydropneumatic tank systems.

However, constant-speed systems have some drawbacks, too. They require large pumps to maintain pressure during peak demand, but during periods of low demand, the pumps still do the same amount of work. This wastes a lot of electricity. Modern energy use guidelines limit the installation of constant speed systems, so review the local code before installing one of these systems.

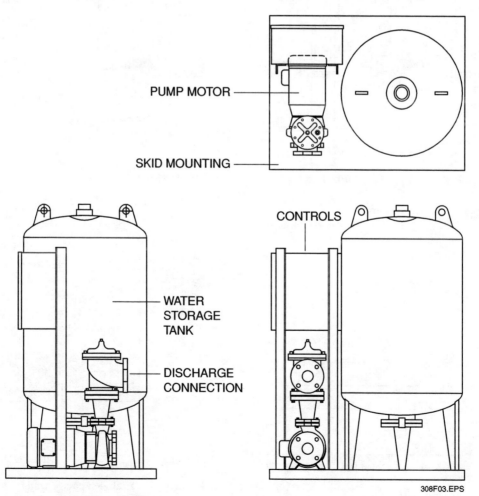

PUMP MOTOR

SKID MOUNTING

CONTROLS

WATER STORAGE TANK

DISCHARGE CONNECTION

308F03.EPS

Figure 3 ◆ Constant-speed system pump and storage tank.

2.1.4 Variable-Capacity Systems

The variable-capacity system was invented to correct the shortcomings of the constant-speed system. Variable-capacity systems use more than one pump (see *Figure 4*). The pumps engage one at a time or all together in response to pressure drops in the system. A plumber programs the pumps to activate at specific times, depending on the anticipated demand. This approach is more energy efficient than having one pump operate all of the time. This makes them more efficient than constant speed systems. Plus, variable-capacity systems don't require a gravity or hydropneumatic tank.

Before a variable-capacity system can be installed, you must establish the water-use cycle for an installation. This can take the form of a chart showing use over time (see *Figure 5*). Then, select pumps that will provide enough water pressure to meet the projected demand at all times. For example, using the chart in *Figure 5*, a plumber might choose to install two pumps. One would provide 33 percent of the system's capacity, and the other would provide 67 percent. During low flow periods, the small pump operates. As demand increases, the second pump engages and the first pump shuts down. To meet the peak evening demand, both pumps operate together (see *Figure 6*).

Some variable capacity systems use a **pressurized air-bladder storage tank** to replenish water in the system. This is a tank in which water pressure is kept constant by compressed air. Pressure builds up in the tank until it reaches the prescribed charge. A flexible bladder or diaphragm keeps the air and water apart. The system's pumps operate only to fill the tank. A pressure switch shuts off the system during periods of no flow (see *Figure 7*).

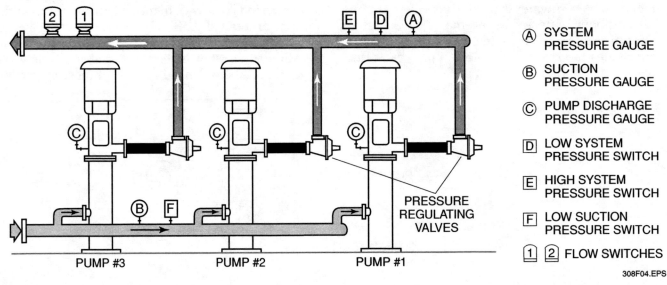

Ⓐ SYSTEM PRESSURE GAUGE

Ⓑ SUCTION PRESSURE GAUGE

Ⓒ PUMP DISCHARGE PRESSURE GAUGE

Ⓓ LOW SYSTEM PRESSURE SWITCH

Ⓔ HIGH SYSTEM PRESSURE SWITCH

Ⓕ LOW SUCTION PRESSURE SWITCH

① ② FLOW SWITCHES

308F04.EPS

Figure 4 ◆ Variable-capacity system.

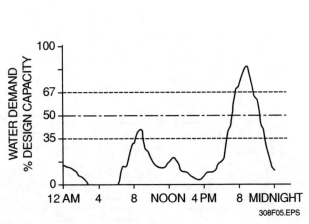

Figure 5 ◆ Water use chart.

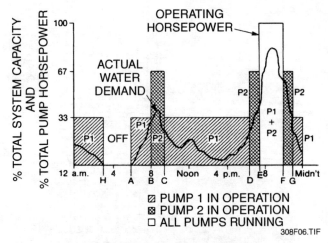

☒ PUMP 1 IN OPERATION
☒ PUMP 2 IN OPERATION
☐ ALL PUMPS RUNNING

308F06.TIF

Figure 6 ◆ Variable-capacity system designed to meet water demand.

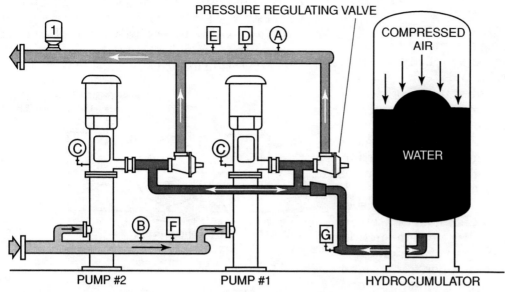

308F07.EPS

Figure 7 ◆ Variable-capacity booster system using a pressurized air-bladder storage tank.

Pressurized tanks have several advantages. The bladder eliminates the threat of corrosion. Because pumps do not have to cycle on and off as often, they last longer. A pressurized tank also eliminates the need for constant pump cycling during periods of no flow caused by minor leaks. A properly designed system with a pressurized tank is highly energy efficient. *Figure 8* illustrates an installation in a multistory building.

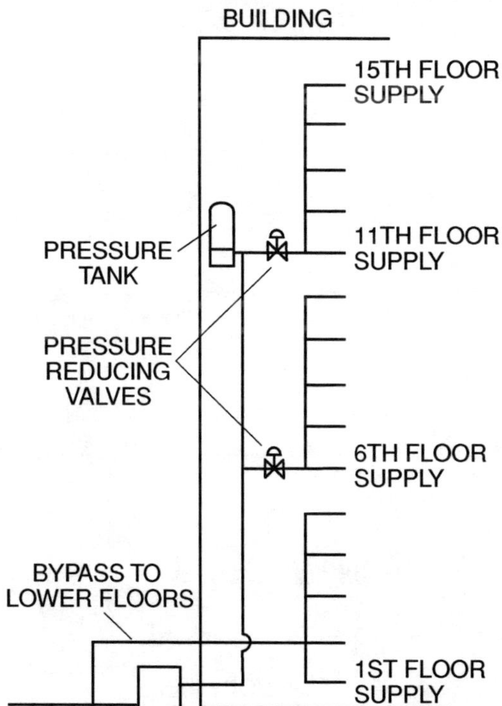

308F08.EPS

Figure 8 ◆ Variable-capacity booster system with an elevated pressurized tank in a multistory structure.

2.2.0 Components of Water Pressure Booster Systems

When designing a water pressure booster system to meet the specific needs of a building or structure, you must consider the intended use, predicted capacity, and the energy requirements. These factors will help determine what components are needed. Components of a water pressure booster system include the following:

- Storage tanks
- Pumps
- Pump motors
- Controls
- Water hammer arresters

Ensure that the components are suitable for the type of water system installation. Refer to the local code for specific requirements governing booster system components.

2.2.1 Storage Tanks

Booster system tanks are manufactured from materials such as galvanized steel, polyethylene, fiberglass reinforced plastic, and even glass bonded to steel. Tanks range in size from 2 gallons to more than 5,000 gallons. They can be purchased from the manufacturer as a kit or built on site. Tank installations may require flexible hoses, manual or automatic float valves, manual shutoff valves, a sight gauge, and a drain. Ensure that the tank materials and accessories are permitted by the local code.

2.2.2 Pumps

Centrifugal pumps are the most widely used type of water pump in plumbing systems (see *Figure 9*) because of their mechanical simplicity and low cost. Centrifugal pumps are most efficient when pumping large amounts of water. Centrifugal pumps can be classified according to how many suction cavities, or intakes, the impeller has. A **single-suction impeller** has a single-suction cavity on one side of the impeller. A **double-suction impeller** has two cavities that are on opposite sides of the impeller. Both single- and double-suction impeller pumps can be used in booster systems. Single-suction impellers work well in low-flow conditions, while double-suction impellers are designed for high-flow conditions.

Single-suction pumps can suffer from a high side load thrust, which is an unbalanced load caused by water exiting only one side of the impeller. Side load thrust can deform the motor shaft over time. Pump seals and impellers will wear out if the condition is allowed to persist. Avoid excess wear of pump seals and impellers by installing a **volute** in the pump body (see *Figure 10*). A volute is a geometrically curved outlet path that directs water from the impeller into the discharge pipe more efficiently than the standard pump outlet design. Volutes improve pump efficiency and reduce wear.

Vertical turbine pumps are also used in water pressure booster systems. They consist of a series of centrifugal impellers stacked one above each other in **stages.** A stage is a single impeller and its water passage. Water is drawn into the first impeller from the bottom. The water is pumped from one stage to the next. The pressure increases at each stage. The pressurized water is finally discharged after it passes through the last impeller (see *Figure 11*). Vertical turbine pumps work well in high-pressure and low-flow conditions, such as in high-rise buildings. They do not experience side load thrust problems.

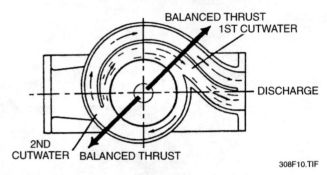

Figure 10 ◆ Booster pump volute.

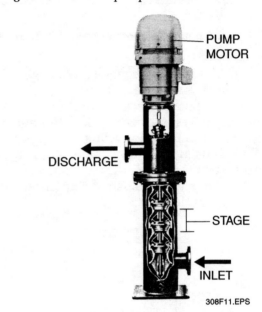

Figure 11 ◆ Vertical turbine pump.

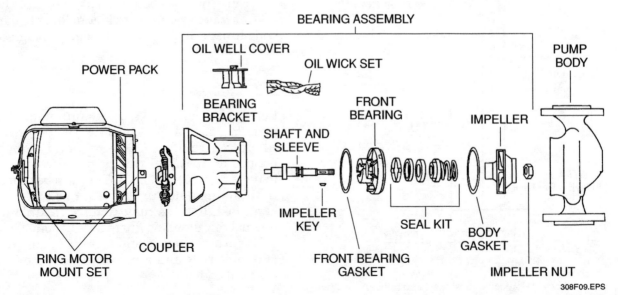

Figure 9 ◆ Centrifugal water booster pump.

2.2.3 Pump Motors

Many booster systems are designed to provide low flow at high pressure. Select a pump that offers the most efficient combination of water flow and pump speed. Note that doubling the pump capacity results in a fourfold increase in head. In other words, a pump that operates at 3,500 revolutions per minute (rpm) can pump twice the capacity of a pump that operates at 1,750 rpm, but the 3,500 rpm pump generates four times the head of the 1,750 rpm pump. Faster pumps operate more efficiently than slower ones.

Faster pumps often operate with more noise than slower pumps. To reduce noise, use a smaller motor with the same capacity. Install **vibration isolators** to cut down on the transfer of vibrations to the structure. These include rubber pads on the pump base and flexible connectors in the inlet and discharge lines.

2.2.4 Controls

Booster system controls include sensors for water pressure, temperature, and flow (see *Figure 12*). A packaged booster system comes with its own controls or you can use other controls if preferred. Use a pressure switch for pumps installed in residential water systems serviced by a well. Install flow sensors in a variable-capacity system to determine which pumps should be functioning. Temperature sensors, also called *thermal sensors*, shut the pumps down during no-flow periods. Temperature sensors detect the heat caused by a spinning impeller after flow stops and shuts off the pump until flow starts again.

308F12.EPS

Figure 12 ◆ Temperature and flow sensors.

2.2.5 Water Hammer Arresters

Water hammer can occur in water pressure booster systems. It is most likely to occur in higher-pressure systems. Water hammer can shorten the service life of the water supply system. Install water hammer arresters to protect the water supply system from damage. They will also help quiet the booster system's operation.

2.3.0 Designing Water Pressure Booster Systems

A water pressure booster system must be selected and designed to meet the needs of the installation. Otherwise, it will not work effectively and efficiently. For example, if the system is intended for a building with **intermittent demand**, such as a hotel or an office building, consider a variable-capacity system. This system is well suited for a wide variation between peak and low demand. A constant-supply system might be appropriate for a hospital, where water is required at all times. The booster system's performance depends on correct estimates of several factors, including the following:

- Energy efficiency
- Type of pump
- Flow rates
- Anticipated water demand

Design the system so that it achieves the best balance of all these considerations. Each of them is discussed in more detail below.

2.3.1 Energy Efficiency

Inefficient booster systems cost more to operate. This consideration is even more important now that energy costs are rising. Inefficient booster systems waste water and electricity. A recent study claims that inefficient booster systems cost building owners $3 million per year in wasted energy. The difference in operational costs between a 10 horsepower (hp) pump motor and a 50 hp pump motor over a one-year period of time could amount to hundreds of dollars. In light of this, be sure to calculate the pumping needs of a system carefully.

Some water pressure booster systems have a **thermal purge valve** installed. These valves drain heated water out of the system when a no-flow condition exists. This procedure wastes large amounts of water, so avoid installing thermal purge valves where operating cost is a concern.

2.3.2 Pump Selection

When selecting the type of pump to use, consider factors such as cost, ease of maintenance, reliability, and operational efficiency (see *Table 1*). Double-suction centrifugal pumps and vertical turbine pumps are very reliable. However, they are also the most expensive types of pump. Keep in mind that the cost of replacement parts will probably be less than the cost of labor needed to install them. Also, consider whether replacement parts for a particular pump are available locally. The cost of shipping replacement parts from across the country may cancel out a pump's low price.

2.3.3 Flow and Demand

Engineers are responsible for calculating flow and demand in a water pressure booster system (see *Table 2*). **Average flow** is considered to be the flow rate under average conditions. This rate cannot be used to determine maximum operating capability of a system, but can be important when calculating off-peak loads. **Maximum flow** is the total flow possible if all outlets are opened at the same time. Because it is highly unlikely that this would happen, this flow factor is unrealistic. The most important factor in sizing water pressure booster systems is **maximum probable flow**. This term refers to the flow under peak demand conditions.

Additional factors that deserve consideration are continuous and intermittent demand. Outlets such as hose bibbs, irrigation pumps, and other relatively constant-flowing devices are considered **continuous-demand** items. When these devices are used, the water flow is constant and makeup water must be at hand to replace it. Intermittent demand, as its name implies, refers to devices such as sinks, lavatories, showers, water closets, and similar devices with usage time of about five minutes or less. Continuous and intermittent demand within a system significantly affect the maximum probable demand factor.

2.4.0 Installing and Maintaining Piping Materials

It is important to make sure that water pressure booster systems are designed and installed correctly. A malfunctioning booster pump usually means an entire water supply system must be shut down until the pump can be repaired. In a home, this can be a nuisance, but in a hospital, it can cost lives. Quick and efficient troubleshooting and repair or replacement is essential.

Table 1 Guidelines for Selecting an Appropriate Booster Pump

| | Single-Suction Centrifugal | | | Double-Suction Centrifugal | Vertical Turbine |
	Close-Mounted	Frame-Mounted	In-Line		
Efficiency	—Especially good in low-capacity range—			Good in high-capacity range	Good in all ranges
Reliability	Good	Good	Good	Best (except at shutoff)	Best
Suction Lift Ability	Good	Good	Good	Best	Best*
Bearings to Maintain	Motor	Motor + 2	Motor	Motor + 2	Motor
Number of Shaft Seals	One	One	One	Two	One
Cost to Repair Seal	High	Moderate	High	Highest	Low
Hours to Repair Seal	2 to 4	1 to 3	2 to 4	4 to 8	1 to 2
Ability to Operate at Shutoff	Poor	Poor	Poor	Poor	Good
Pump Curve	Flat	Flat	Flat	Flat**	Steep**
Motor Availability	Fair	Good	Poor***	Good	Fair
Space Required	Least	Medium	Least	Most	Least
Original Price	Low	Moderate	Moderate	High	High

Notes:
   * Can also tank mount pumps to eliminate suction lift.
   ** Multistage pumps can be selected further to left or right on curve.
   *** If specially designed shaft is used.

Table 2 Sample Table for Calculating Booster System Flow and Demand

Design Data			
System Capacity			_____ GPM
Pressure Required at Discharge Header			_____ PSIG
Pressure Drop Through Package (include discharge valve)			_____ PSIG
Minimum Suction Pressure			_____ PSIG

Pump Data	Pump No. 1	Pump No. 2	Pump No. 3 (Optional)
Gallons per Minute	_____	_____	_____
Pump TDH (feet)	_____	_____	_____
Header size _____ inch			
Discharge Valve:			
❏ PRV or ❏ Check Valve Size	_____	_____	_____
Series _____ Size	_____	_____	_____
Motor HP _____			
Motor RPM _____ Voltage _____		Hertz _____	Phase _____

Configuration	Pump Orientation	Header Material
	❏ Vertical	❏ Copper
	❏ Horizontal	❏ Galvanized Steel

Ensure that the pipes and connections are suitable for use in a potable water system. Refer to your local code. Pipes must not corrode or degrade as a result of contact with potable water. Local codes also specify limits on lead content in pipes. Do not use any of the following joints or connections in a water supply:

- Cement or concrete
- Nonapproved fittings
- Solvent joints between different types of plastic pipe
- Saddle-type fittings

Permitted pipe materials include acrylonitrile-butadiene-styrene (ABS), polybutylene (PB), cross-linked polyethylene (PEX), polyvinyl chloride (PVC), and chlorinated polyvinyl chloride (CPVC) plastic; asbestos-cement; brass; gray and ductile iron; copper; and steel. Use mechanical joints when connecting pipes made of different materials. Local codes will provide specific guidance.

2.5.0 Installing and Maintaining Water Tanks

Depending on the design, the plumber may be responsible for locating the water tank. Be sure to provide access for tank maintenance and tank removal. When designing a booster system for a multistory building, remember that installing the tank higher allows a greater head. This not only increases the service pressure but reduces the pumping requirement.

Provide adequate support for the tank. Ensure that all underground tanks are covered. This will prevent unauthorized access, contamination, and infestation. Do not locate tanks or manholes for potable water pressure tanks underneath soil or waste piping or any other possible source of contamination. Install a valved drainpipe at the lowest point of each tank (see *Table 3*). Also, provide a vacuum relief valve at the top of the tank. Pressurized air-bladder storage tanks may not require a relief valve. On hydropneumatic systems, install a pressure relief valve on the supply pipe. Ensure that the relief valve is equal to the tank's rating.

Table 3 Size of Drainpipes for Water Tanks

Tank Capacity (Gallons)	Drainpipe (Inches)
Up to 750	1
751 to 1,500	1½
1,501 to 3,000	2
3,001 to 5,000	2½
5,001 to 7,500	3
More than 7,500	4

Source: International Code Council, Inc. See Figure and Table Credits, page 8.24.

Leaks in a pressure tank can cause excessive pump cycling and low water pressure. Test for leaks by applying a mixture of soap and water to the exterior of the tank above the water line. Bubbles will appear where air is escaping from the tank. Repair the leak using methods approved for potable water systems.

2.6.0 Installing and Maintaining Motors and Pumps

Install a low-pressure cutoff on booster system pumps to prevent back siphonage caused by negative pressure. Prevent tank overflow by installing ball cocks on inlet piping. Follow the manufacturer's installation instructions when installing pumps. Ensure the pump is primed before testing, if required.

Knowing how to diagnose pump problems quickly and accurately will not only save you time, but will also save the owner money. Whether testing a new pump or repairing an existing one, the most common problems include the following:

- Pump will not start
- Pump motor has overheated
- Pump cycles excessively
- Pump will not turn off
- Pump delivers little or no water

Review the troubleshooting tips for each type of problem before answering a service call. The more experience you have in the field, the more familiar you will become with pump problems and solutions. Ask other experienced plumbers for advice. Some problems may be more common in your area due to climate and local codes.

2.6.1 Problems With the Pump Starting

If a pump motor will not start, examine the pump's electrical system. If there is a blown fuse or if the motor is short-circuiting, contact an electrician to secure or replace the wiring. This work should be done only by a licensed electrician. Mechanical problems may also keep a pump motor from starting. Examine the booster system controls. Reset improperly adjusted temperature, pressure, and flow switches. Ensure that the tubing to the pressure switch is not plugged. Test the impeller by first turning the pump off, then attempting to turn the impeller with a screwdriver. If the impeller does not rotate, it may be blocked. Remove the pump casing and inspect the impeller for blockage.

2.6.2 Problems With the Pump Motor Overheating

Pump motors have overload protectors to trip the motor off in the event of overheating. If a motor overheats, use a voltmeter to test the line voltage. If the voltage is below the required level, contact an electrician to ensure that the wiring from the electrical main to the motor is the correct size. The electrician should also ensure that the motor is wired according to the manufacturer's instructions. Again, only those licensed to correct electrical problems should attempt to do so.

High air temperature and inadequate venting can also cause overheating. Ensure that the pump is adequately vented and that the surrounding air temperature is within the pump's specified limits. Another cause of overheating is prolonged operation at low water pressure. If this is the case, install a globe valve on the water discharge line and throttle it to provide the desired pressure.

2.6.3 Problems With the Pump Cycling Excessively

Excessive cycling may be caused by leaks in various locations in the system. Seals are the most frequently required repairs for domestic water pumps. Inspect all seals for leaks and replace if necessary. To test for leaks on the discharge side, shut off all fixtures in the system. Inspect the discharge piping while listening for the sound of running water. The sound of running water will indicate a leak. Check all valves, especially ball cocks, for leaks. Repair or replace any valves that

 DID YOU KNOW?

Fresh Water Supply in America's First Modern Hotels

Boston's Tremont Hotel opened in 1829. It has been called America's first modern hotel. The four-story building featured metal bathtubs in the basement and indoor toilets on the ground floor—an unheard-of luxury in those days. The water for these fixtures came from a metal water tank on the roof. A steam engine pumped the water from ground level to the tank. Five years later, the Tremont's designer, Isaiah Rogers, was commissioned to build the six-story Astor House in New York City. This hotel had 300 guestrooms. Rogers equipped 17 rooms on the upper floors with water closets and baths for the guests. The Tremont Hotel and Astor House were the first large modern buildings in the United States equipped from the outset with their own plumbing systems.

WARNING!

You must follow special arrangements when immersing pump motors in water. When doing this type of work, you are at risk of electrocution, which can cause serious injury and even death. Before working on a pump motor, ensure that the pump motor is turned off and power to the pump motor is disconnected. Dry the pump motor thoroughly before operating. Wear appropriate personal protective equipment, and follow proper installation and maintenance procedures. Sometimes you have only one chance to do it right!

are leaking. Use a pressure gauge to find a leak on the suction side of a shallow well system. To identify a leak on the suction side of a deep well system, use the following steps:

Step 1 Attach a pressure gauge to the pump.

Step 2 Close the discharge line valve.

Step 3 Using a compressor or hand pump, apply a pressure of about 30 pounds per square inch (psi) to the system.

Step 4 Once this pressure has been reached, stop pumping and read the pressure gauge. If the pressure drops, there is a leak somewhere in the suction side. Inspect all pipes and valves.

2.6.4 Problems With the Pump Turning Off

If the pump will not turn off, check the pressure switch. It may be set too high for the actual operating conditions, or it may have "drifted" to an incorrect setting. An electrician can inspect the pressure switch for defects caused by arcing in the electrical system. Clear the pressure switch tube. Ensure that the pump is correctly primed. Inspect the pump ejector and remove any blockage.

2.6.5 Problems With the Pump Delivering Little or No Water

Electrical and mechanical problems can cause a pump to deliver little or no water. Only a licensed electrician can check the line voltage and the priming. To clear an air lock, turn off the pump for about one minute and then restart it. Repeat this several times until the air is cleared from the pump. Ensure that the piping to the pump is correctly sized and free of leaks. Inspect the air volume control for leaks, and repair or replace the control, if required. Check the setting of pressure regulating valves against the manufacturer's

instructions. Ensure that the pump ejector is correctly sized. Remove any blockages in the intake, impeller, and ejector. Inspect the pump for worn parts and install appropriate replacements.

Review Questions

Sections 1.0.0–2.0.0

1. Before installing a variable capacity system, establish the _____ in the building.
   a. size of the fixtures
   b. water use cycle
   c. amount of electricity consumed
   d. location of the water tank

2. In a pressurized air-bladder storage tank, water pressure is kept constant by _____.
   a. compressed air
   b. balancing valves
   c. thermal purge valves
   d. system pressure

3. Maximum probable flow refers to water flow under _____ conditions.
   a. system design
   b. average demand
   c. total possible flow
   d. peak demand

4. Installing a low-pressure cutoff on a booster pump will prevent _____.
   a. back siphonage
   b. overflow
   c. back pressure
   d. short circuits

5. _____ are the parts in domestic water pumps that most frequently require repair.
   a. Floats
   b. Impellers
   c. Seals
   d. Mounts

3.0.0 ◆ RECIRCULATION SYSTEMS

When you take a shower at home, you may have to let the water run for a little while before it gets hot. However, in a hotel, the water is usually hot as soon as you turn on the tap. This is because the hotel uses a **recirculation system**. This is a special installation that constantly circulates hot water throughout a building's water supply piping. Recirculation systems can provide hot water instantaneously to any fixture in the system.

Recirculation systems work best in larger buildings, such as hotels, apartments, schools, factories, and restaurants. They are less effective in buildings where the farthest hot water tap is less than 100 feet from the water heater. Most recirculation systems are designed by building engineers. However, plumbers need to know about the different types of recirculation systems, their components and layouts, and the basics of installation.

3.1.0 Types of Recirculation Systems

There are three types of recirculation systems. In an **upfeed system,** hot water is supplied to fixtures as the water travels up the system (see *Figure 13*). In a **downfeed system,** hot water is supplied as the water travels down the system (see *Figure 14*). In a **combined upfeed and downfeed system,** hot water is supplied on both the upward flow and downward flow (see *Figure 15*). If required by the design, each of these systems may include hot water storage tanks. In all three

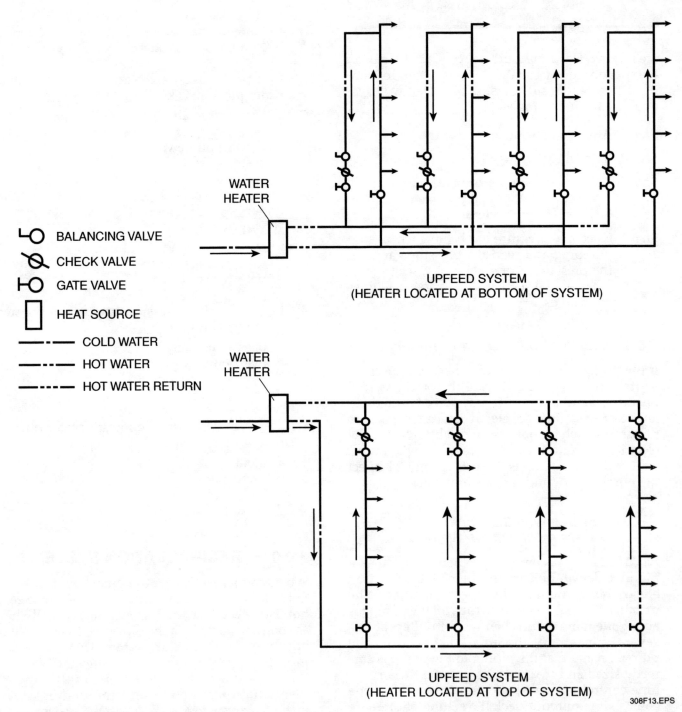

BALANCING VALVE

CHECK VALVE

GATE VALVE

HEAT SOURCE

— · — COLD WATER

— · · — HOT WATER

— · · · — HOT WATER RETURN

WATER HEATER

UPFEED SYSTEM
(HEATER LOCATED AT BOTTOM OF SYSTEM)

WATER HEATER

UPFEED SYSTEM
(HEATER LOCATED AT TOP OF SYSTEM)

308F13.EPS

Figure 13 ◆ Upfeed recirculation systems.

types of recirculation systems, the water heater and storage tank can be located at either the top or the bottom of the installation.

When you are selecting a recirculation system, there are several things you should consider. These include whether it is an upfeed or a downfeed system and whether the water heater is positioned at the top or the bottom of the building. The various ways these systems can be configured require more or less piping. For example, as shown in *Figure 13*, an upfeed system with the water heater on top requires less piping than an upfeed system with the water heater on the bottom. Conversely, in a downfeed system, as shown in *Figure 14*, less piping is required when the water heater is located at the bottom, and more piping is required when the water heater is located at the top. Regardless of which system you select, be aware that the potential for leaks increases according to the amount of pipe used, not according to the location of the water heater.

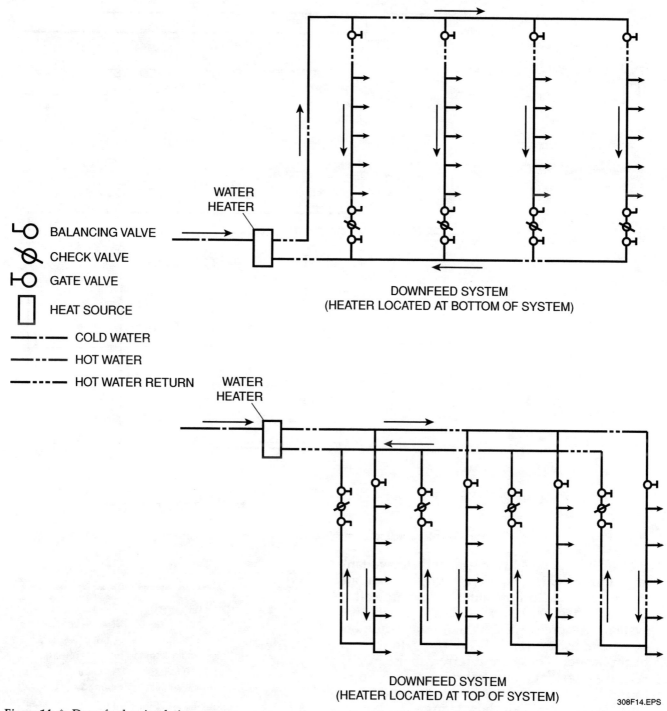

Figure 14 ◆ Downfeed recirculation systems.

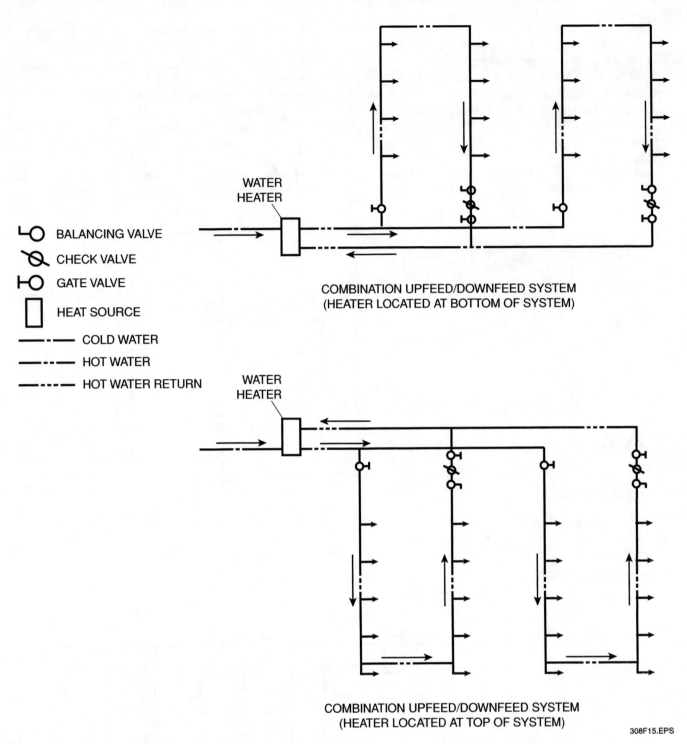

BALANCING VALVE

CHECK VALVE

GATE VALVE

HEAT SOURCE

—— · —— COLD WATER

—— · · —— HOT WATER

—— · · · —— HOT WATER RETURN

WATER HEATER

COMBINATION UPFEED/DOWNFEED SYSTEM
(HEATER LOCATED AT BOTTOM OF SYSTEM)

WATER HEATER

COMBINATION UPFEED/DOWNFEED SYSTEM
(HEATER LOCATED AT TOP OF SYSTEM)

308F15.EPS

Figure 15 ◆ Combined upfeed and downfeed recirculation systems.

Systems that have the heater and storage tank at the top usually require a pump to circulate the water back to the heater. These are called **forced circulation systems.** Today, most hot water systems use forced circulation. A smaller number of systems use the force of gravity to return water to the heater. These are referred to as **gravity return systems.** Gravity return systems can be found mostly in smaller and older buildings. Plumbers need to know how to install and maintain both kinds of hot water recirculation systems.

WARNING!

Do not install valves on T/P relief lines. Valves on a relief line could cause the line to block. Blockage could damage the water heater or even cause it to explode. A malfunctioning water heater can result in water damage to property. An explosion can cause injury and even death.

3.1.1 Laying Out a Gravity Return System

Gravity return systems are fairly small systems that have the water heater and storage tanks installed below the system. That way, water can flow back into the heater without mechanical help (see *Figure 16*). Water in a gravity return system moves more slowly than water in a forced circulation system. However, the water flow is constant. Because of these facts, gravity return systems are used only in smaller buildings where speed is not as important. They can also be found in older buildings that have not been upgraded.

Proper grade is important in a gravity return system. Locate the supply pipe so that it either inclines upward or rises straight up to the top of the building. That way, the cold water forces the hot water up into the system. Return pipes should also have an appropriate degree of slope. Upfeed systems use a sloped supply pipe, while downfeed systems use a single vertical hot water riser. The single vertical hot water riser permits hot water to rise without interference. It is the most efficient arrangement. Grade is usually specified in construction drawings or in the local code. If not, get the information from the local plumbing inspector.

Circulation in a gravity system depends on the difference between the weight of cold and hot water. In a system with more than one circulating loop, ensure that the differential is the same in each loop. Use balancing valves to ensure that there is no differential in the system. Otherwise the system will not work as efficiently.

3.1.2 Laying Out a Forced Circulation System

Forced circulation can be used in both upfeed and downfeed systems, but it is most commonly used where the heater and tank are above the system. This design is often used in newer and taller buildings. Forced circulation can also be used in buildings with long horizontal runs of pipe, such as schools. The layout is essentially the same as that of a gravity return system. The main difference is the addition of a pump to circulate the hot water (see *Figure 17*). Because forced circulation systems do not rely on the force of gravity to move the water, you will need to select a pump that is rated for the system's operating conditions. Install the pump on the return line before the water heater. Provide a gate valve on the inlet and outlet sides of the pump, and install a check valve on the inlet pipe after the gate valve.

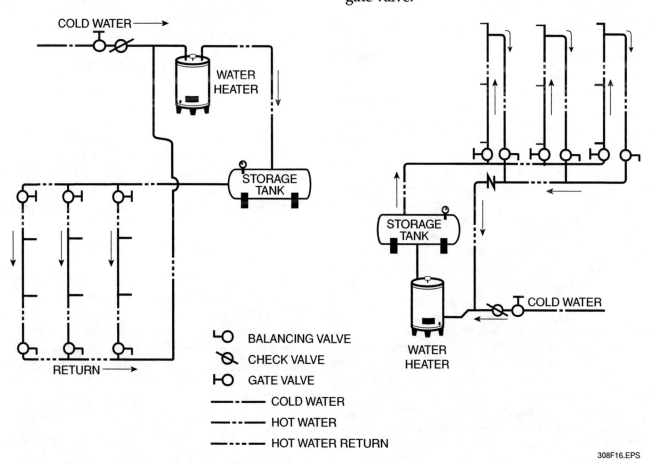

308F16.EPS

Figure 16 ◆ Gravity return in downfeed and upfeed systems.

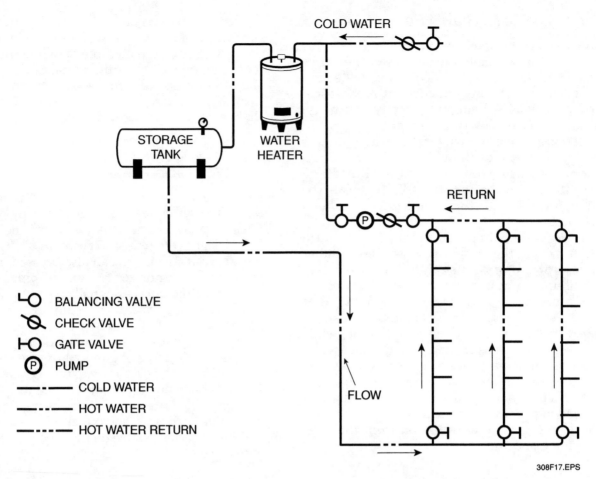

BALANCING VALVE

CHECK VALVE

GATE VALVE

(P) PUMP

—··— COLD WATER

—··— HOT WATER

—··— HOT WATER RETURN

308F17.EPS

Figure 17 ◆ Forced circulation in an upfeed system.

DID YOU KNOW?

Hot Water in Ancient Rome

Long before the development of recirculation systems, ancient Rome used hot water in its public and private baths. Public baths served as community centers where people gathered to hear news, talk with friends, and relax. The bathhouses were huge, ornate structures. The indoor baths of the Emperor Diocletian, for example, could seat 3,000 people. Most bathhouses had separate baths with cold, warm, and hot water. Water for the baths was heated by an underground furnace that also heated the air in the baths. The walls were covered in mosaic tiles. Water poured out from silver spigots shaped like fish and mythical creatures.

To supply the baths, Roman engineers accomplished some of the most complex engineering feats of their day. Water was channeled to the city's baths through 220 miles of gravity-powered aqueducts and tunnels. At its height, this network provided about 300 gallons of water for each citizen per day. By 400 C.E., Rome had 11 public baths, 1,352 public fountains and cisterns, and 856 private baths.

3.1.3 Multiple Heaters and Storage Tanks

Large recirculation systems often require more than one heater or storage tank. The number of units is determined by the anticipated hot water demand. The building engineer determines the need for multiple heaters and tanks. Designers can combine heaters and storage tanks in several ways. There are several basic guidelines for designing systems with multiple heaters and storage tanks, including the following:

- Do not allow hot water to circulate through a cold heater.
- Arrange piping and valves so that the heaters and tanks can function independently or together.
- Use as few valves as possible.
- On long pipe runs from heaters to storage tanks, ensure that the pipe runs have the proper degree of slope.
- Connect two water heaters to the discharge pipe using 45-degree wyes in the direction of flow.

These guidelines are general recommendations only. Consult the plumbing drawings for specific details.

3.2.0 Components of Recirculation Systems

In addition to water heaters, hot water recirculation systems use storage tanks, pumps, valves, and temperature controls to manage the flow of hot water. Standard pipes and fittings carry the water. Building engineers choose the components that suit the needs of each installation. In this section, you will learn about the components that make up a recirculation system. Consult your local code for guidelines that are specific to your area.

3.2.1 Storage Tanks

Some systems use automatic storage tank water heaters (see *Figure 18*), which heat and store water at a controlled temperature and deliver it on demand. Other systems require a separate storage tank. Storage tanks can be used with most electric and gas fired water heaters. Do not install storage tanks in systems that use indirect, also called *instantaneous*, water heaters. Storage tanks must be pressurized, so be sure to install pressure relief valves as required. The size of the storage tanks depends on the size of the facility. The engineers will estimate the size of the tank. Plumbers should consult the construction drawing to determine the size and refer to the local code for specific requirements.

308F18.EPS

Figure 18 ◆ Hot water storage tank.

3.2.2 Pumps

Forced circulation systems use centrifugal pumps to circulate water through the hot water lines. Use forced circulation when water cannot circulate by gravity. Pumps increase the efficiency of a recirculation system. They can be designed to operate one

WARNING!
The maximum operating pressure of a pump is listed on its nameplate. Do not exceed this pressure, or the system could fail. This could not only damage the system, but could also cause injury or death.

of two ways. One option is to have the pump on constantly, which allows the water to circulate constantly through the system. The other option is to operate the pump by a temperature control, which allows the pump to switch on when the water cools below a preset minimum temperature. The choice depends on the type of installation and the demand for hot water. Some codes require that pumps be shut off when the system is not operating.

Use stainless steel or bronze pumps to handle potable water. These metals do not rust, which would contaminate water and may clog pipes. Consult local codes for additional requirements about recirculation pumps.

3.2.3 Pipes

Piping materials used in hot water recirculation systems are generally the same as those that are used with cold water. Building engineers specify their size and type. Codes require that hot water pipes withstand a given pressure at high temperatures. Select only valves and joints that are made from compatible materials, and ensure that the lead content of the pipe material does not exceed local standards. Pipes should be properly insulated in order to help make the system more energy efficient. Follow the building plans closely when installing pipe. Undersized or oversized pipes will cause the system to malfunction.

High water temperatures will cause pipes and joints to expand. Install **expansion joints** in hot water piping as directed by the building plans. These will allow the pipes to expand and flex when hot water flows through them. Expansion joints can be installed several ways. Consult your local code. The building plan may call for expansion loops made from runs of pipe. The building plans will provide the specifications.

ON THE
· LEVEL ·

A recent estimate shows that homes can waste anywhere from 27 to more than 100 gallons a day waiting for running water to get hot. That means an efficient hot water recirculation system can save between 9,800 and 38,000 gallons per year in just one home. Depending on the installation, a recirculation system could represent significant cost savings to the customer and reduced wastewater burdens in the community.

3.2.4 Valves

You may already be familiar with a type of **tempering mixing valve.** They are used in single-handle faucets. Hot water recirculation systems also use tempering valves (see *Figure 19*). They control water temperature by adding cold water to the hot water flow. This ensures that the water in the system will remain at the right temperature. Install tempering mixing valves in systems that use indirect heaters.

Check valves are automatic valves that permit liquids to flow in only one direction. They protect against backflow caused by back pressure and back siphonage. Check valves are widely used in a variety of plumbing installations. They come in horizontal and vertical styles (see *Figure 20*). You learned how to install and service check valves in *Plumbing Level Two.*

 DID YOU KNOW?

Balancing Hot and Cold Water

In 1964, Powers Regulator Company (now Powers Process Controls) patented a way to use water pressure to protect bathers from sudden changes in water temperature. The Hydroguard™ 410 Pressure Balancing Valve had a pressure-equalizing chamber operated by a diaphragm. The diaphragm could detect a drop in water pressure caused by a sudden loss of either cold or hot water. People taking showers no longer had to fear being scalded if the cold water suddenly failed, nor did they have to risk being hit with an icy surprise first thing in the morning.

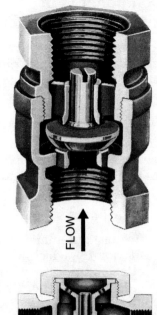

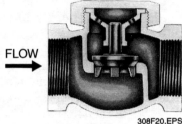

308F20.EPS

Figure 20 ◆ Check valves.

Install either a tap or an air relief valve at the highest point in the system in order to allow air in the pipes to be purged. Adjustable **balancing valves** at the base of each water line allow the plumber to find and set the most efficient flow. Balancing valves compensate for pipe friction and other irregularities, ensuring a smooth and steady flow.

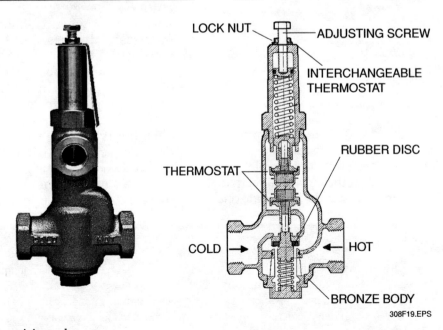

308F19.EPS

Figure 19 ◆ Tempering mixing valve.

3.2.5 Aquastats

An aquastat is a thermostat that regulates the temperature of hot water in a boiler (see *Figure 21*). It can be used to control the water heater's operation. Install aquastats on or near the water heater or at high points in the water lines. **High-limit aquastats** are a special kind of aquastat. They turn off the boiler when the water inside reaches maximum temperature. These aquastats act as safety devices.

308F21.EPS

Figure 21 ◆ Aquastats.

3.2.6 Expansion Pipes and Tanks

Some designs require the installation of **expansion pipes** or **expansion tanks.** An expansion pipe is a section of pipe in the hot water supply that is wider than the rest of the system piping. It allows overheated water to expand as it enters. The expansion cools and slows down the hot water. Cold water in the pipe helps the process further. Expansion tanks are attached to the hot water supply pipes and perform the same role as expansion pipes.

 WARNING!

Hot water under pressure can escape with explosive force, and it can cause dangerous scalding. At 125°F, it takes about 1½ to 2 minutes for hot water to cause scalding. At 155°F, it takes about 1 second. Wear appropriate personal protective equipment. Review and follow all construction drawings closely.

Install expansion pipes and tanks above the system where the hottest water collects. Review the building plans to determine this location. Use pipes or a combination of pipes and a tank to allow the desired expansion. Expansion pipes and tanks can be installed in both gravity and forced circulation systems.

3.3.0 Installing and Maintaining Recirculation Systems

Plumbers install and repair hot water recirculation systems. Many of the materials and the methods used to install recirculation systems are common to other types of plumbing installations. You will need to read the construction drawings before constructing the system. Your knowledge and experience will be of help when you build a recirculation system, but you also need to be aware of some problems that are unique to recirculation systems. They are discussed below.

3.3.1 Installing a Hot Water Recirculation System

Depending on the design of the recirculation system, you may need to connect more than one riser to the hot water supply main. Ensure that the connection to each riser is correct. Otherwise, problems may develop with the hot water flow in some of the risers. Incorrect connections in a gravity system can keep it from working right. Correct connections are also important in systems that have been altered by previous repairs. Each riser may require a slightly different connection because of the connection's location and the riser's length and slope. Four different connections can be used:

- 45-degree connection (the most typical)
- Horizontal connection
- Vertical connection
- Inverted 45-degree connection

Knowing when to install each type of connection comes from practical experience. For example, a short riser might need a vertical connection to draw as much water as a longer one with a 45-degree connection. The longest riser might need a horizontal connection so that it will not draw more water than the others. Consult your local code for guidelines, and also seek advice from experienced plumbers.

Millions of Solar Roofs

New city, state, and federal government programs are encouraging homeowners and businesses to use solar hot water heating. Solar energy is a renewable resource and creates few greenhouse gas byproducts. Since 1992, the federal government has issued tax credits to businesses that use solar power. At least 30 states offer similar incentives for homeowners and businesses. The United States Environmental Protection Agency (EPA) has launched a program called the Million Solar Roof Initiative. The goal is to have solar power installations on a million private homes by 2010. The federal government also intends to convert 20,000 federal buildings to solar power by that time.

Some systems convert sunlight into electricity. Others collect solar energy and use it to heat fresh water supplies and swimming pools directly. These direct systems, called *solar thermal systems*, are the least expensive form of solar energy for applications connected to power grids. Solar power is gaining popularity as people become more environmentally conscious. Plumbers can expect to install more and more solar hot water systems in the near future.

When installing more than one water heater or storage tank in a high-rise building, install check valves on the return risers. These valves will prevent the backflow of cold water into the hot water supply system. Install check valves near the water heater to prevent hot water from reentering the cold water system. Check your local codes for specifications. Tempering mixing valves can be located in several places in the system. One effective location is outside the heat exchanger on the discharge line.

Install centrifugal pumps on the hot water return line. Locate them as close as possible to the water heater (refer back to *Figure 17*). Install pumps with a bypass or a union. In some high-rise buildings, it may be necessary to install several pumps on different floors. If so, follow the plans and specifications closely.

3.3.2 Troubleshooting and Maintaining a Recirculation System

Recirculation systems are made up of components that you have already learned about in your professional apprenticeship training. You have learned how to troubleshoot problems with water tanks, water heaters, pipes, and valves. Therefore, by drawing on the knowledge you have already gained from earlier modules and your past experiences, you will be able to troubleshoot a recirculation system. In some cases, a manufacturer's service technician may be required to do this work.

Other things to consider include the use of balancing valves to prevent pressure differentials between the stacks in a recirculation system with more than one stack. Balancing valves compensate for different lengths and heights of pipe in various parts of the system. Open and close the individual valves until the pressure is uniform throughout the system.

Scale buildup is another common occurrence in a recirculation system. Remember that scale buildup is caused by deposits. These deposits can create problems, such as clogged pipes. Clogged pipes will need to be cleaned or replaced.

Review Questions

Section 3.0.0

1. In forced circulation systems, the heater and storage tank are located _____ the system.
   a. at the bottom of
   b. in front of
   c. at the top of
   d. adjacent to

2. Connect two water heaters to the discharge pipe using _____ in the direction of flow.
   a. 45-degree wyes
   b. sanitary tees
   c. 45-degree ells
   d. double tapped tee-branches

3. To purge air in the recirculation system pipes, install a(n) _____ at the highest point in the system.
   a. tempering mixing valve
   b. check valve
   c. air relief valve
   d. aquastat

4. Install expansion pipes and tanks _____ the recirculation system.
   a. in the upfeed risers of
   b. at the top of
   c. at the bottom of
   d. in the downfeed risers of

5. Connect pumps to the hot water return line using a bypass or a _____.
   a. union
   b. flange
   c. compression ring
   d. reducer coupling

Summary

Both water pressure boosters and hot water recirculation systems use techniques and tools that are familiar to plumbers. However, each system must be designed to fit the building or structure. They must be able to provide hot and cold water at the correct pressures and temperatures. Plumbers design, install, and maintain booster systems. Building engineers usually design hot water recirculation systems, but plumbers install and maintain these systems.

Water pressure booster systems increase water pressure to the required level. The four types of booster systems are elevated tank, hydropneumatic tank, constant speed, and variable capacity. Consider several factors when designing a booster system including the intended use, the predicted capacity, and the energy requirements. Pumps, motors, and tanks require periodic maintenance.

Recirculation systems distribute hot water throughout a building. They work best in larger buildings. There are three types of recirculation systems, called *upfeed*, *downfeed*, and *combined upfeed and downfeed systems*. These systems can be installed with water heaters and storage tanks above or below the system. Systems with heaters and tanks at the top, called *forced circulation systems*, usually require a pump to circulate the water. These systems are used in buildings with long horizontal runs of pipe, such as a school. Systems with heaters and tanks at the bottom rely on gravity to circulate the water. They are called *gravity return systems*. Read the construction drawings before installing a recirculation system. Always ensure that water systems will perform the way they were designed.

Trade Terms Introduced in This Module

Average flow: The flow rate in a *water pressure booster system* under average operating conditions, used to calculate off-peak loads.

Balancing valves: Valves in a *recirculation system* that ensure steady water flow.

Combined upfeed and downfeed system: A *recirculation system* in which hot water is supplied in the upward and downward flow through the system.

Constant-speed system: A *water pressure booster system* in which water circulates continuously through the system and pumps replacement water from the main.

Continuous demand: Demand for water caused by outlets, pumps, and other devices with a relatively constant flow.

Double-suction impeller: A pump impeller used in *water pressure booster systems* with two cavities on opposite sides of the impeller. It is used to channel water in high-flow conditions.

Downfeed system: A *recirculation system* that supplies hot water as it travels down the system.

Elevated tank system: The oldest and simplest type of *water pressure booster system*, featuring a water tank on the roof operated by the weight of the water column.

Expansion joint: A mechanical device installed on a pipe that allows it to expand and flex as hot water flows through it.

Expansion pipe: A section of pipe in a *recirculation system*, wider than the regular pipe, that allows water to cool and slow down as it flows.

Expansion tank: A tank attached to the supply pipes in a *recirculation system* that allows water to cool and slow down as it flows.

Forced circulation system: A *recirculation system* that uses a pump to circulate hot water through the system.

Gravity return system: A *recirculation system* in which the force of gravity circulates hot water throughout the system.

High-limit aquastat: A safety device that deactivates a hot water boiler when the water reaches the maximum temperature.

Hydropneumatic tank system: A *water pressure booster system* that uses air pressure to circulate the water.

Intermittent demand: Demand for water caused by fixtures that are used no more than about five minutes at a time.

Maximum flow: The theoretical total flow in a *water pressure booster system* if all outlets are simultaneously opened.

Maximum probable flow: The flow in a *water pressure booster system* during periods of peak demand.

Pressurized air-bladder storage tank: A tank used in a *water pressure booster system* that uses compressed air to maintain water pressure.

Recirculation system: A plumbing installation that circulates hot water within a building, providing customers with hot water on demand.

Single-suction impeller: A type of impeller used in centrifugal pumps that have a single-suction cavity. Single-suction impellers are used in low-flow *water pressure booster systems*.

Stage: In a *vertical turbine pump*, the arrangement of a single impeller and its water passage.

Tempering mixing valve: A valve that adds cold water to the hot water flow to control water temperature.

Thermal purge valve: A valve that allows heated water to drain from the *recirculation system* when no-flow conditions exist.

Upfeed system: A type of *recirculation system* that supplies hot water as it travels up the system.

Variable-capacity system: A type of *water pressure booster system* that uses two or more pumps to provide water in response to peaks and lows in demand.

Vertical turbine pump: A pump used in *water pressure booster systems* where high-pressure and low-flow conditions exist.

Vibration isolators: Devices such as rubber pads and flexible line connectors that reduce the effects of pump vibration.

Volute: In centrifugal pumps, a geometrically curved outlet path.

Water pressure booster system: A plumbing installation that increases water pressure in the fresh water supply system.

Additional Resources

Efficient Building Design Series, Volume 3: Water and Plumbing, 1999. Ifte Choudhury and J. Trost. New York, NY: Prentice Hall.

Plumbing, Cold Water Supplies, Drainage, and Sanitation, 1981. Fred Hall. New York, NY: Van Nostrand Reinhold.

Plumbing, Hot Water Supply, and Heating Systems, 1981. Fred Hall. New York, NY: Van Nostrand Reinhold.

Acknowledgments

Charles Owenby, J. F. Ingram State Technical College
ITT Bell & Gossett

Figure and Table Credits

Bell & Gossett	308F09
Canariis Corporation	308F03
Fairbanks Company	308F20
Pacific Plumbing Company	308F05, 308F06, 308F12
Parker Boiler Co.	308F18
Synchro Flo, Inc.	308F04, 308F07, 308F10, 308F11
Taco, Inc.	308F21
Watts Regulator Company	308F19

Table 3: Written permission to reproduce this material was sought from, and granted by, the copyright holder, **International Code Council, Inc.,** 5203 Leesburg Pike, Suite 600, Falls Church, VA 22041.

References

1997 International Plumbing Code, 1997. Falls Church, VA: International Code Council, Inc.

Grundfos Pump Corporation Web site, www.us.grundfos.com, "Hot Water Recirculation: Frequently Asked Questions," www.us.grundfos.com/web/prodmisc.nsf/pages/hwr-main-menu, reviewed August 2000.

Planning Drain, Waste & Vent Systems, 1993. Howard C. Massey. Carlsbad, VA: Craftsman Book Company.

NCCER CRAFT TRAINING USER UPDATES

The NCCER makes every effort to keep these textbooks up-to-date and free of technical errors. We appreciate your help in this process. If you have an idea for improving this textbook, or if you find an error, a typographical mistake, or an inaccuracy in the NCCER's Craft Training textbooks, please write us, using this form or a photocopy. Be sure to include the exact module number, page number, a detailed description, and the correction, if applicable. Your input will be brought to the attention of the Technical Review Committee. Thank you for your assistance.

Instructors – If you found that additional materials were necessary in order to teach this module effectively, please let us know so that we may include them in the Equipment and Materials list in the Instructor's Guide.

Write: Curriculum Revision and Development Department
National Center for Construction Education and Research
P.O. Box 141104, Gainesville, FL 32614-1104

Fax: 352-334-0932

E-mail: curriculum@nccer.org

Craft	Module Name
Copyright Date	Module Number	Page Number(s)
Description		

(Optional) Correction

(Optional) Your Name and Address

Servicing Piping Systems, Fixtures, and Appliances

COURSE MAP

This course map shows all of the modules in the third level of the Plumbing curriculum. The suggested training order begins at the bottom and proceeds up. Skill levels increase as you advance on the course map. The local Training Program Sponsor may adjust the training order.

PLUMBING LEVEL THREE

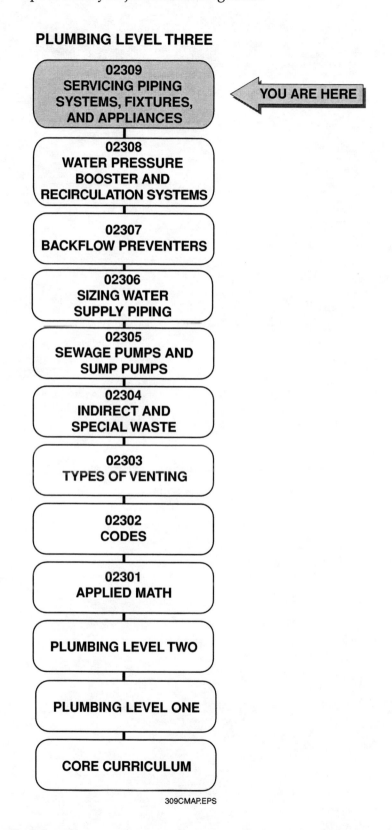

02309
SERVICING PIPING
SYSTEMS, FIXTURES,
AND APPLIANCES

YOU ARE HERE

02308
WATER PRESSURE
BOOSTER AND
RECIRCULATION SYSTEMS

02307
BACKFLOW PREVENTERS

02306
SIZING WATER
SUPPLY PIPING

02305
SEWAGE PUMPS AND
SUMP PUMPS

02304
INDIRECT AND
SPECIAL WASTE

02303
TYPES OF VENTING

02302
CODES

02301
APPLIED MATH

PLUMBING LEVEL TWO

PLUMBING LEVEL ONE

CORE CURRICULUM

309CMAP.EPS

Figures

Table

Servicing Piping Systems, Fixtures, and Appliances

Objectives

When you have completed this module, you will be able to do the following tasks in accordance with local codes:

1. Diagnose water supply problems.
2. Diagnose water quality problems.
3. Explain different types of corrosion and their effects on pipes.
4. Diagnose and solve fixture and appliance problems.
5. Troubleshoot and repair water supply problems.
6. Troubleshoot and repair water heater problems.
7. Troubleshoot and repair water drainage problems.
8. Troubleshoot lawn irrigation systems.

Prerequisites

Before you begin this module, it is recommended that you successfully complete the following: Core Curriculum; Plumbing Level One; Plumbing Level Two; Plumbing Level Three, Modules 02301 through 02308.

Required Trainee Materials

1. Appropriate personal protective equipment
2. Pencil and paper
3. Copy of your local code

1.0.0 ◆ INTRODUCTION

Plumbing systems require periodic maintenance. Sometimes, poor water quality can cause problems. Roots may block drainpipes. A pipe can degrade and leak. A fitting may break. Fixtures can simply wear out from constant use. Plumbers decide whether to repair or replace worn or broken items. They need to be able to diagnose and fix problems quickly and correctly. Finding the cause of a problem is a lot like detective work. Always take the time to look for the cause of the problem. Choose the best possible solution, not just the fastest or the least expensive. These steps will reduce or eliminate callbacks and result in more satisfied customers. This means more business in the future.

Fixing plumbing problems takes a lot of knowledge and many years of experience, but you have already learned many of the basic skills that you will need to perform the work according to professional standards. Beginning plumbers can develop their knowledge and experience in several ways. One way is to read plumbing literature, such as manuals, trade journals, and codes. Pay special attention to articles that offer practical tips. Another way is to pay attention when working on plumbing systems. Notice how the various parts are installed. Think about why piping and fixtures were installed a certain way. Finally, work with experienced plumbers. Watch how they diagnose and fix problems. Do not be afraid to ask a lot of questions. The best plumbers are the ones who never stop learning.

2.0.0 ◆ GENERAL GUIDELINES FOR SERVICE CALLS

Before you can begin any repair work, you should learn and follow some general service and safety guidelines. It is impossible to predict what you will find when responding to a service call. For example, a small leak from a faucet spout into a kitchen sink, while wasteful and annoying, is not an emergency. Usually, it's also not hazardous. On the other hand, a leaking water heater is potentially dangerous. In an emergency like this, you must stop the flow of water, immediately minimize the potential damage, and make the area safe before you begin the repairs. Keep your work area as neat as possible. Do not put a customer's property at risk by hurrying to finish a job.

2.1.0 Service

Plumbers pride themselves on their customer service skills as well as their technical knowledge. Four guidelines for success are:

- Promptness
- Preparedness
- Thoroughness
- Courtesy

Customers appreciate plumbers who observe these guidelines. When you make an appointment, keep it. Show up for service calls on time. If you cannot make your appointment or if you are going to be late, notify the customer as soon as you can. These are courtesies you would expect from someone coming to your own home.

WARNING!
Do not work in a crawl space or other restricted area without prior confined space training. Toxic and flammable vapors can accumulate in confined spaces, as well as carbon monoxide (CO) emissions. Electrical, hot water, and steam lines also pose risks in confined spaces. Wear appropriate personal protective equipment, and wear a dust mask or respirator where needed.

WARNING!
Electrical shocks don't happen just to electricians! Always protect yourself against accidental contact with energized sources when working on plumbing installations and responding to service calls.

Arrive prepared to do the job. Know before you arrive what work needs to be done on site. Ensure that the proper tools and supplies are available. Customers lose confidence in plumbers who have to leave and return several times just to complete a simple task.

While working at the customer's home or workplace, keep the work area as clean as possible. Lay down protective coverings over floors, furniture, and equipment. Wear overshoes and gloves while working. These will help keep dirt and grease off the walls, floors, and furniture. When possible, remove furniture and other obstacles from the work area. Do the work neatly and clean up when the job is done. These simple, commonsense steps will help ensure satisfied customers and future business.

2.2.0 Safety

The following are some general safety guidelines for you to follow when responding to a service call. Adapt these guidelines to the job at hand. As you progress in your career, you'll add your own guidelines from your experience.

- Wear rubber-soled shoes or boots for protection from slipping and electric shock.
- Turn off electrical circuits.
- Shut off a valve upstream from the leak. If you think it will take you a while to find and turn off this valve, direct the leak into a suitably sized container to minimize damage until you can turn off the water.
- Remove excess water.
- Place ladders carefully, bracing them if necessary.
- When finished, turn the water back on and test your repair.

ON THE · LEVEL ·

Cleaning Up is Part of the Job

Whether you're installing water supply piping or repairing a fixture, you must always clean up your work site. A clean work site is safer for you and for any trades that come in after you. Cleaning up is especially important if you are called in to repair a major leak. No one wants to pay for repairs and then have to replace furnishings or carpet because you were careless.

3.0.0 ◆ SERVICING WATER SUPPLY SYSTEMS

You have learned that customers have a right to expect adequate fresh water on demand. Water supply systems meet that need. Water supply systems are made of many components such as pipe, fittings, valves, traps, and fixtures. Larger systems also use pumps and storage tanks to circulate the water efficiently. Parts can and do wear out over time. They may stop working efficiently, or they may break altogether. When either happens, the water system cannot provide fresh water on demand. In some cases, a malfunction can cause illness or even death. Plumbers are responsible for repairing or replacing broken or defective components. They are also responsible for making sure the repaired system operates as smoothly and efficiently as before.

3.1.0 Leaks

Leaks are the most common problem that plumbers encounter in water supply systems. Leaking and broken pipe can be hazardous. Pipes installed in a new building can break when the building shifts and settles. Underground pipes can break if a water line trench is improperly backfilled. Leaks in a new system are usually caused by improper installation or damaged components. In older systems, the problem is often a worn component or a corroded supply line.

Before beginning work, find out how old the building is and determine what materials were used for the piping and fittings. This information offers clues to the nature of the problem. A leak indicates a problem with one or more components in the plumbing system. Common causes of leaks include:

- Physical damage to pipes
- Improper connections
- Corroded pipes
- Frozen pipes
- Excessive vibration or water hammer
- Higher than normal system pressure

Locating a leak is often difficult. Leaks can occur at a fitting or along the length of pipe. Water can run along a pipe or structure, showing up far from its source. Take the time to find the location of a leak by tracing the water drip or stain to its source. Fixing a leak without correcting the problem that caused it will only result in repeated plumbing problems. Such a situation could cause the customer to lose confidence in you.

 DID YOU KNOW?
Proper Flushing

After any repair to a water supply system, always purge the system when the repairs are complete. Remove all screens and filters and allow the water to run through the lines for a reasonable length of time. This step will ensure that all foreign matter is flushed out of the system. Advise customers to have water heaters and tanks flushed periodically to remove sediment and mineral deposits. Manufacturers generally recommend flushing five gallons of water from hot water tanks each month for a reasonable amount of time. Make sure that the drain is fully opened. This will allow proper flushing.

3.1.1 Repairing a Leak Along a Fitting

Use good repair practices to repair a leak near a fitting in plastic pipe. Make sure the pipe is dry before trying to repair it. Cut out the fitting using a pipe cutter or other cutting tool. A mini tube cutter is also a useful tool for cutting pipe (see *Figure 1*). Test fit the new fitting. When the fit is correct, use couplings to join the new fitting to the existing pipe (see *Figure 2*). To repair a leak in a copper fitting, solder the fitting.

309F01.EPS

Figure 1 ◆ Mini tube cutter.

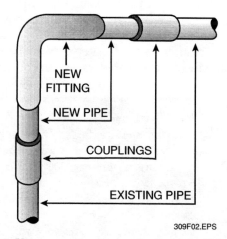

309F02.EPS

Figure 2 ◆ Using couplings to repair a leak at a fitting.

3.1.2 Repairing a Leak Along a Length of Pipe

To repair a leak in a length of pipe, use good repair practices. Make temporary repairs in plastic and copper pipe using pipe clamps (see *Figure 3*), then join them together using the skills you learned in *Plumbing Level Two*.

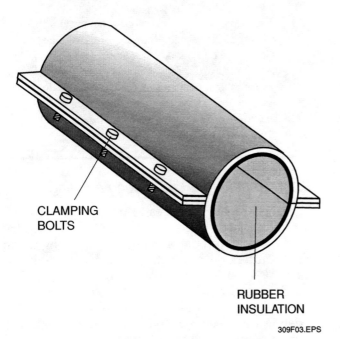

CLAMPING
BOLTS

RUBBER
INSULATION

309F03.EPS

Figure 3 ◆ Temporary pipe clamp.

3.1.3 Eliminating Pipe Condensation

Sometimes, water stains will appear on walls or ceilings where there are no leaking pipes. These stains are often caused by **condensation** and roof leaks. Condensation is moisture that collects on cold surfaces that are exposed to warm, moist air. Dripping condensation causes water stains and musty mildew odors. To eliminate condensation, add insulation to the pipe (see *Figure 4*). Insulation prevents the cold pipe from coming into contact with warm air. It is the most effective way to protect against condensation. If adding insulation is impractical, increase the ventilation. Doing so will reduce the amount of moisture in the air.

WARNING!

Slips, trips, and falls cause most general industry accidents, according to the Occupational Safety and Health Administration (OSHA). They cause 15 percent of all accidental deaths, and they are second only to motor vehicles as a cause of fatalities. Slips, trips, and falls happen quickly, so take a few minutes to make your work area safe.

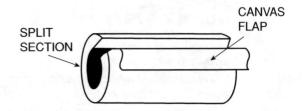

SPLIT
SECTION

CANVAS
FLAP

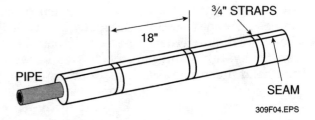

¾" STRAPS

18"

PIPE

SEAM

309F04.EPS

Figure 4 ◆ Insulating a pipe.

3.2.0 Frozen Pipes

Water supply piping must be protected from freezing. If the piping has frozen, first determine whether the line has ruptured. If a water supply pipe has ruptured, turn off the water supply before attempting to thaw the line. Install temporary clamps to seal minor leaks. If the pipe is severely damaged, replace the damaged section immediately.

There are several ways to thaw frozen pipes safely. An electric thawing machine uses high amperage over a long period. Its direct current melts the ice from the walls of the pipe, allowing water to flow freely through the line. A hot water thawing machine uses hot water to thaw frozen pipe. Although it does not work as quickly as an electric thawing machine, it is less expensive. Whichever method you use, always follow the manufacturer's specifications.

When the frozen water line is exposed, an electric heat gun may be used. This is ideal when only a portion of the line is frozen. Directing hot air over the frozen line thaws the line. The running water can thaw the remainder of the line. If the frozen supply line is undamaged, a heat lamp or hair dryer can also be used to thaw it. If the pipe is indoors, heating the air can help. Also consider putting a rag over the pipe and pouring hot water over the rag. Do not use boiling water, which can cause the pipes to expand—and possibly burst—if they are excessively heated.

Repair freeze-damaged pipes with the same techniques used for fixing a leak. Note that ice usually forces the pipe to expand. As a result, repair fittings may not attach easily. Therefore, you must cut the pipe well beyond any areas that have frozen.

Tips for Preventing Frozen Pipes

The less frequently a water pipe is used, the more likely it is to freeze. Here are some easy steps to prevent frozen pipes that you can suggest to homeowners:

- Allow the water to run slowly throughout the night or during the coldest periods.
- Drain outside hose connections at the beginning of the cold-weather season.
- Insulate exposed water pipes.
- Use heat tape to protect the water line from freezing.
- Close gaps that allow cold air into spaces where water pipes are located.

Where freezing is a problem, install approved frost-free **yard hydrants** (see *Figure 5*) and **wall hydrants**. Consult with local experts to make sure the hydrant has the necessary approvals. A yard hydrant is a fresh-water spigot connected to a pipe that runs directly to the water main. Yard hydrants work by having their valve located below the frost line. If yard hydrants continue to freeze, ensure that the water supply pipe has the proper pitch. The excess water will then be able to drain from piping when the valve is closed.

As the name suggests, a wall hydrant is installed in a building's exterior wall. Wall hydrants work by draining water after the valve is closed. The valve seat is located on the interior side of the building's wall. Wall hydrants are sized according to the thickness of the building's wall.

CAUTION

Underground supply lines can freeze if they are installed where snow is not allowed to accumulate. These areas include driveways, sidewalks, and patios. Supply lines beneath these areas are subject to freezing because there is no snow to insulate them. Consult the local code and building officials to identify the frost line in your region.

Wrap pipes with **electric heat tape** for temporary protection in case of frost. Electric heat tape is made of flexible copper wires coated in a protective layer of rubber or other nonconductive material. When wrapped around a pipe and plugged into a wall socket, heat tape provides enough heat to keep pipes from freezing. Be aware of the danger of melting plastic pipe. Install heat tape according to the manufacturer's instructions. Do not wrap heat tape over itself. Heat tape can be wrapped under insulation.

To prevent frozen pipes, also consider placing the pipe in partitions. When working with outdoor faucets, install frost-proof hose bibbs or install a drain or waste cock inside the building in a heated area.

WARNING!

Never use an open flame to thaw a pipe. Torches, lighters, and other sources of open flame are dangerous and could cause a structural fire. Using a torch on a frozen water line may damage solder joints, create hazardous steam pockets, and ignite surrounding building materials.

3.3.0 Water Pressure

Water supply systems are usually designed to operate at pressures between 35 to 50 pounds per square inch (psi). You learned about water supply systems in the module *Sizing Water Supply Piping*.

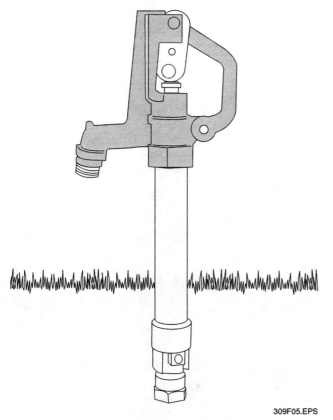

309F05.EPS

Figure 5 ◆ Yard hydrant.

Pressure below or above the normal range can cause problems with the water supply. Too little pressure means fixtures and outlets cannot work efficiently. Too much pressure causes noise, water hammer, and excessive wear.

The allowable pressure range for noncirculating water, called **static water pressure**, is usually between 15 and 75 psi. The upper limit is the maximum setting for pressure-reducing valves (PRVs). Static pressure should be kept below 75 psi. Static pressure tends to be higher for commercial systems with flushometers, while pressure tends to be lower for residential systems. Fifty psi is a good average pressure for homes connected to public water supply systems. Homes connected to private water supply systems operate closer to 40 psi.

The system pressure may fluctuate daily during peak and low-use periods. Fluctuations also may be seasonal. Water consumption is usually much higher in the summer than in the winter. The pressure range for circulating water, also called **dynamic water pressure**, is lower than that of static water pressure. Remember to distinguish between static and dynamic water pressure when designing a water supply system. Refer to your local code.

3.3.1 Low Pressure

Problems caused by low water pressure are generally harder to fix than those caused by high pressure. They also tend to be more expensive. Low pressure is caused by problems with either static pressure or dynamic pressure. Low pressure in the city main is a common static pressure problem. Common dynamic pressure problems include the following:

- A defective water meter
- A leak in a building main
- A defective or partly closed valve
- Improperly sized pipe
- **Rust,** corrosion, kinks, and other pipe blockages

Customers often mistake deposit buildups for low water pressure. Mineral buildups and corrosion in pipes restrict water flow. The result is a dynamic pressure drop in the system. This is often a problem in older areas of a municipal water supply system. You may need to conduct a pressure test of the system to confirm the diagnosis. If the problem is extensive, a new water supply system will have to be installed.

Remember that low pressure can cause backflow if there is a cross-connection in the system. For example, opening a valve on the ground floor of a multistory building could cause back siphonage in a fixture several stories up. Low-pressure problems should be corrected quickly to prevent this.

Install a pressure booster system to correct low pressure. Otherwise, the potable (drinkable) water supply could become contaminated.

Too many fixtures on a supply line can also cause low water pressure. In such cases, replace the supply lines to handle the additional demand. Refer to the local code before replacing any water supply pipe. Be sure to select the correct size pipe for the anticipated demand. Remember to reconnect the water service to the house.

3.3.2 High Pressure

High pressure can cause vibrations and knocking in water pipes. High pressure in the water main is one of the causes of water hammer. (A quick-closing valve in the system is the other common cause of water hammer.) Simply anchoring the pipe to a wall or frame may not solve the problem. Install water hammer arresters instead (see *Figure 6*). Air chambers can also correct vibrations caused by high pressure (see *Figure 7*).

Another way to correct water hammer vibrations is to install pressure-reducing valves (see *Figure 8*) at the service entrance to a structure or near the source of the high-pressure problem. Install shutoff valves on both the inlet and outlet sides of PRVs (see *Figure 9*). Installing shutoff valves on both sides allows water to flow through the bypass for access to the PRV for maintenance and repair.

Note that the excess pressure may be occurring in only one part of the system. For example, a system in a high-rise building could have high pressure at a branch line near the service entrance and low pressure at the lowest end. If this is the case, do not place the PRV in the main service line.

309F06.EPS

Figure 6 ◆ Water hammer arrester.

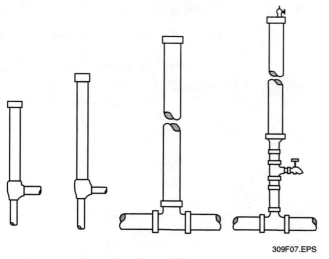

Figure 7 ◆ Examples of air chambers.

309F07.EPS

Doing so will reduce the pressure at the most distant outlet. The entire piping system may have to be redesigned as a result of the excess pressure. Each specific water hammer problem requires its own unique solution. Discuss various problems and solutions with experienced plumbers. Refer to the local code as well.

High water pressure is also a common cause of whistling noises in supply lines. Pipes and fixture openings that are too small can also cause whistling. Take the time to diagnose the cause of the whistling before attempting to correct the problem. Pressure-reducing devices are an effective way to eliminate a whistling noise. Reducing the pressure in the problem area is usually easier than replacing the piping.

If the whistling is limited to a small area, check the pipe for restrictions such as a burr that was not properly reamed. Check for defective fittings, malfunctioning valve washers, and loose valve stems. If the problem persists, install a pressure-reducing device in the branch line. If the whistling occurs throughout the entire structure, install the pressure-reducing device at the service entrance.

3.3.3 Bladder Tanks

Private water supply systems that draw on wells or springs usually store water in bladder tanks (see *Figure 10*). Bladder tanks provide adequate pressure for the supply system. First, ensure that the tank's pressure switch is properly adjusted. Pressure switches can be set to different pressure ranges, so select a range that is appropriate for the building's water pressure needs. Most switches have settings for 20 to 40 pounds, 30 to 50 pounds, and 40 to 60 pounds. Consult the building's plumbing drawings to determine the pressure requirements.

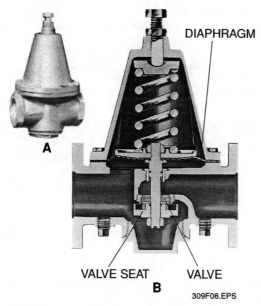

Figure 8 ◆ Pressure-reducing valve.

DIAPHRAGM

A

VALVE SEAT   VALVE

B

309F08.EPS

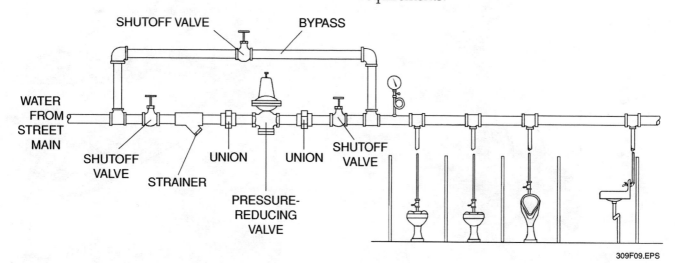

Figure 9 ◆ Installation of a pressure-reducing valve.

SHUTOFF VALVE   BYPASS

WATER FROM STREET MAIN

SHUTOFF VALVE   STRAINER   UNION   UNION   PRESSURE-REDUCING VALVE   SHUTOFF VALVE

309F09.EPS

Figure 10 ◆ Bladder tank for a private water supply system.

A tank with low pressure cannot provide adequate water pressure throughout the system. If this is the problem, you need to pressurize the bladder tank. Begin by finding the correct pressure setting in the manufacturer's specifications. The correct pressure switch setting is important because it determines when the pump is activated. First, empty the tank completely. Then recharge the tank to the correct pressure, using the tank's **Schrader valve** (see *Figure 11*). Schrader valves, which are usually located at the top of the tank, are small valve stems used to pump air into the tank. Test whether the tank can maintain a pressure range of 20 pounds. If it cannot, the chamber may have lost air pressure. Inspect the air chamber and repair any leaks that you find.

3.4.0 Water Quality

An important part of a plumber's job is to ensure that a building's water service meets quality standards. Water quality problems can affect the health of customers. These problems can also damage the water supply system. Therefore, plumbers must be able to correct water quality problems quickly and efficiently.

When water quality may be a problem, start by testing the water. Test results will help you to determine the right purification technique. Choosing a purifier without testing may not solve the problem, and it also could force customers to spend much more money than necessary. Research the problem up front. It takes a lot less time and effort in the long run.

Two common causes of water quality problems are hard water deposits and suspended or dissolved particles. Each of these is discussed in more detail in the following sections.

3.4.1 Hard Water Deposits

Look inside an old coffeepot or teakettle. You likely will see a rough, grayish coating on the bottom. These deposits are mineral salts such as calcium and magnesium. Heat causes the mineral salts to settle out of water, called *precipitating*, and stick to the metal surface. Water with large amounts of minerals is called **hard water**. Hard water deposits, also called *scale*, can collect in hot water supply pipes and water heaters (see *Figure 12*). Hard water deposits can block pipes, reducing flow. Pipes with excessive mineral salt buildup must be replaced. Mineral salts also collect on heating elements in water heaters. This buildup makes water heaters work less efficiently.

Figure 11 ◆ Schrader valve.

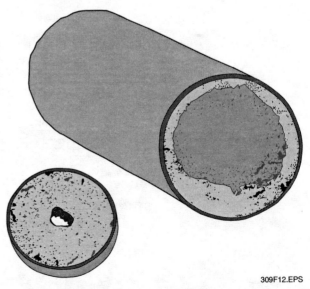

Figure 12 ◆ Scale buildup in a water main.

Plumbers usually diagnose hard water problems using a water test. Treat hard water using a **water softener** (see *Figure 13*). A water softener is a device attached to the water supply line that chemically removes mineral salts from the water before they can enter the system.

One popular type of softener is the **zeolite system**. Zeolite systems consist of a resin tank, also called a *mineral tank*, and a brine tank. Hard water enters the mineral tank. The calcium and magnesium salts in the water flow through resin in the tank. The resin is saturated with sodium, which reacts chemically with the salts and draws them out of the water. The minerals are then washed out the resin tank's drain. When the cycle is complete, sodium-rich water from the brine tank recharges the resin tank. (Water supply treatment will be covered in more detail in *Plumbing Level Four*.) Remember that some codes require that specialists service water softeners.

Installing a water softening system is a fairly big, costly procedure. Always discuss specific needs with the customer before installing a water softener. People may have specific requirements.

For example, some people cannot drink softened water for medical reasons.

In a home, the amount of water that needs to be softened may vary. For example, hose bibbs and lawn sprinkler outlets don't need softening. If the water is slightly hard, then only the hot water system will need to be softened. If the water is very hard, install softeners for the entire interior system (see *Figure 14*).

Mineral deposits can also occur in systems where the water is not very hard. High heat can cause these mineral deposits. The water heater settings may be too high. The solution may simply be to lower the water heater thermostat setting, which will help keep the water below the critical temperature above which mineral deposits form. Thermostats are extremely sensitive. Therefore, adjust the thermostat in small increments. When you have made the first small change in the thermostat setting, show the customer how to make further changes. This extra step can eliminate callbacks. If the problem persists, you will have to evaluate the entire system's hot water needs.

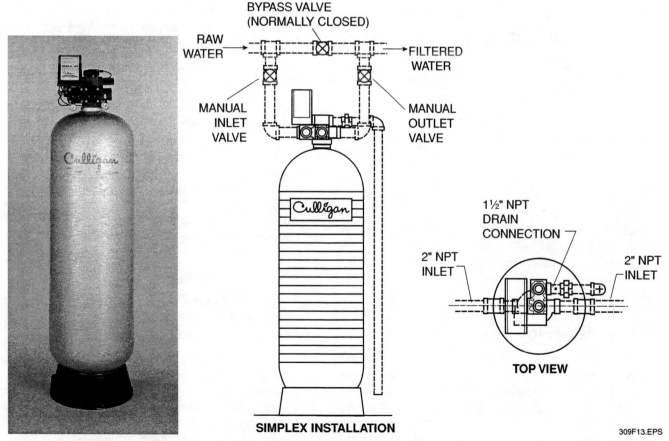

Figure 13 ◆ Water softener.

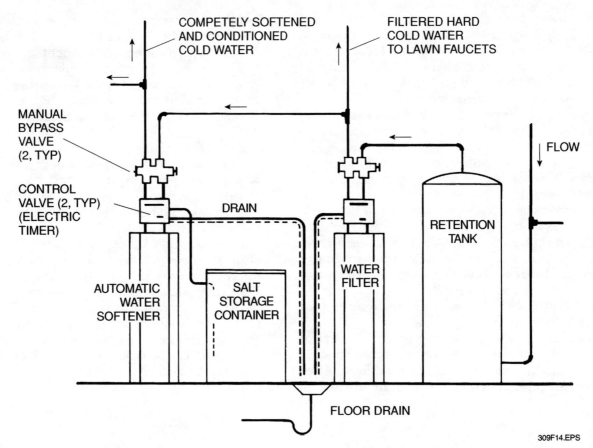

COMPLETELY SOFTENED AND CONDITIONED COLD WATER

FILTERED HARD COLD WATER TO LAWN FAUCETS

MANUAL BYPASS VALVE (2, TYP)

CONTROL VALVE (2, TYP) (ELECTRIC TIMER)

DRAIN

FLOW

RETENTION TANK

AUTOMATIC WATER SOFTENER

SALT STORAGE CONTAINER

WATER FILTER

FLOOR DRAIN

309F14.EPS

Figure 14 ◆ Typical water softening installation.

3.4.2 Small Particles and Organisms

Testing the water will indicate whether there are small particles and organisms in the water. These particles and organisms can lower water quality. Install a **water filter** to remove such matter from potable water (see *Figure 15*). There are many different types of filters, each designed to trap certain types of materials. Be sure to consult the manufacturer's information when selecting a filter. Many filters use cellulose acetate membranes to separate out impurities. This is called the **reverse osmosis method**. To remove sulfur from water, install an **activated carbon filter**. This filter is filled with charcoal. The charcoal absorbs sulfur from the water as the water passes through the filter. Sediment filters trap dirt and silt. Install a **deionizer** on the cold water inlet to trap metals in the water. Deionizers also filter out mineral salts.

3.5.0 Water Flow Rate

Earlier in this level you learned how to calculate flow in a water supply system. The flow rate is the volume of circulated water in a plumbing system. It is calculated in gallons per minute (gpm). The rate of flow in a water supply system can be controlled by flow restrictors. In many cases, these devices can increase system effi-

ciency by reducing water and energy consumption. For example, a flow restrictor installed in a typical shower can cut water consumption nearly in half without affecting the shower's performance.

Install flow restrictors at the inlet lines of individual fixtures that operate at a lower pressure than that of the system. That way, individual fixtures can have different flow rates. Follow the manufacturer's instructions when installing flow restrictors.

Many water supply systems rely on pumps to control the flow rate. These pumps may break down and need to be replaced. Systems that use pumps include the following:

• Wells
• Irrigation systems
• Hot water recirculation systems
• Pressure booster systems
• Solar water heaters

In the module titled *Sewage Pumps and Sump Pumps*, you learned how to install and service pumps. When a pump breaks down, the entire water supply system usually has to be shut down. Plumbers must diagnose and correct pump problems quickly and efficiently.

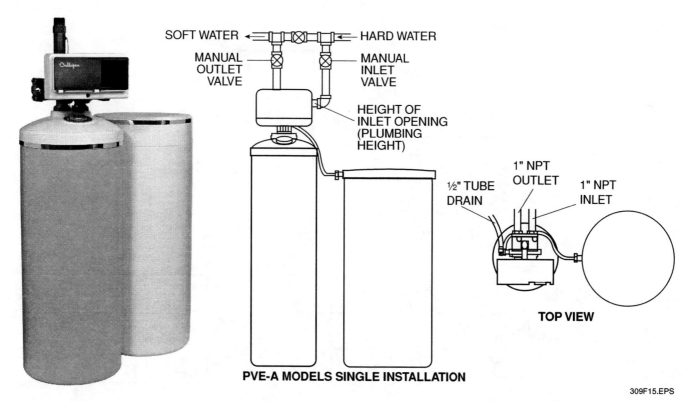

SOFT WATER ◄— HARD WATER
MANUAL OUTLET VALVE
MANUAL INLET VALVE
HEIGHT OF INLET OPENING (PLUMBING HEIGHT)

½" TUBE DRAIN
1" NPT OUTLET
1" NPT INLET

TOP VIEW

PVE-A MODELS SINGLE INSTALLATION

309F15.EPS

Figure 15 ◆ Water filter for commercial applications.

When replacing a pump, ensure that the replacement pump is the proper size. The new pump should be able to handle the same amount of water and use the same amount of electricity as the old pump. If a pump breaks down repeatedly, review its specifications. Compare that information with the system's flow rate calculations and plumbing drawings. Ensure that the pump is correctly matched with the system. A building engineer may need to conduct the appraisal.

Use gate and check valves when installing a pump (see *Figure 16*). Install the pump using either a union or flange connection so that it can be removed easily. Some water supply systems rely on more than one pump to operate (see *Figure 17*). Ensure that the system can operate on one pump while the other is being repaired or replaced.

Some fixtures must be filled with water to a specific level to do their job. Water closet tanks and washing machines are two common examples. These fixtures do not require flow restrictors. However, plumbers can adjust the water levels in these fixtures. Such adjustments allow the fixture to operate efficiently and use less water. Consult the manufacturer's installation instructions, and local codes may also provide guidance. Experience is a plumber's best guide.

ON THE
· LEVEL ·

The Safe Drinking Water Act

Congress passed the Safe Drinking Water Act in 1974 and revised it in 1986. The Act is the main federal law that ensures the quality of drinking water in the United States. The Act requires municipal water authorities to test and report on the quality of the water regularly. They must test the amount of minerals, organic matter, and bacteria in the water. The government can fine water authorities if the water in their area does not meet the Act's standards for purity. For this reason, many water authorities have installed backflow preventers to protect the potable water supply.

You can read the complete text of the Act online at Cornell Law School's Legal Information Institute at http://www4.law.cornell.edu/uscode/42/300f.html.

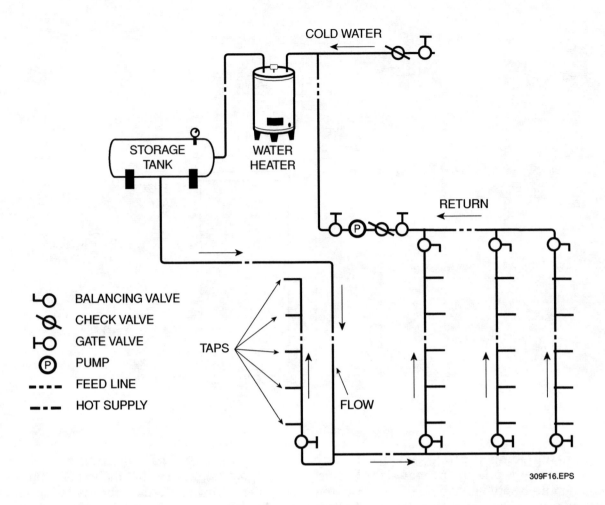

COLD WATER

STORAGE TANK

WATER HEATER

RETURN

⊶ BALANCING VALVE
⊘ CHECK VALVE
⊶ GATE VALVE
Ⓟ PUMP
---- FEED LINE
---·- HOT SUPPLY

TAPS

FLOW

309F16.EPS

Figure 16 ◆ Typical pump installation in a hot water recirculation system.

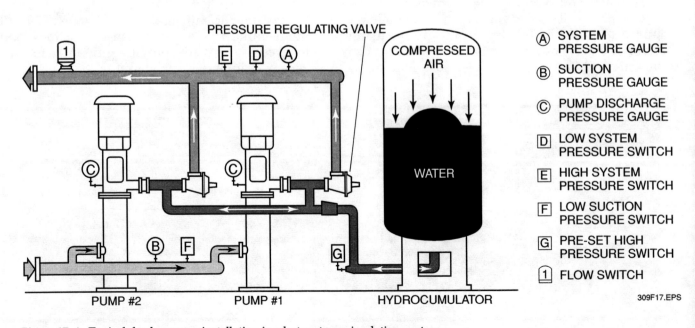

PRESSURE REGULATING VALVE

E D A

COMPRESSED AIR

WATER

Ⓐ SYSTEM PRESSURE GAUGE

Ⓑ SUCTION PRESSURE GAUGE

Ⓒ PUMP DISCHARGE PRESSURE GAUGE

Ⓓ LOW SYSTEM PRESSURE SWITCH

Ⓔ HIGH SYSTEM PRESSURE SWITCH

Ⓕ LOW SUCTION PRESSURE SWITCH

Ⓖ PRE-SET HIGH PRESSURE SWITCH

① FLOW SWITCH

PUMP #2   PUMP #1   HYDROCUMULATOR

309F17.EPS

Figure 17 ◆ Typical duplex pump installation in a hot water recirculation system.

3.6.0 Water Heaters

The water heater is one of the most expensive appliances in a home. Water heaters must be selected, installed, adjusted, and maintained with care. When a water heater leaks or fails to supply enough hot water, it may need to be replaced. When there is no hot water at the tap, the water heater is the most likely cause. Common water heater problems include:

- Failure of the pilot light or heating element
- Improper thermostat setting
- Uninsulated hot water lines
- Overuse of hot water
- Improper size

Solutions for each of these problems are discussed in the following section.

3.6.1 Correcting Common Water Heater Problems

Failures of the pilot light and the heater element are the most common water heater problems. If there is no hot water, begin by checking for an extinguished pilot light or if the electric element is burned out.

Gas water heaters burn a mix of air and fuel gas to heat water (see *Figure 18*). For a gas water heater with an extinguished pilot light, check for drafty conditions. Redirect louvers to the equipment

room so that air does not blow directly on the pilot light. If that does not correct the problem, install additional shielding around the water heater. Shielding will protect the pilot light from drafts. Check for dust and cobwebs blocking the jet. This is especially important if the pilot light has been off for a long time. If the gas jet is blocked, clean it according to the manufacturer's instructions. Inspect the vent and clear any blockages.

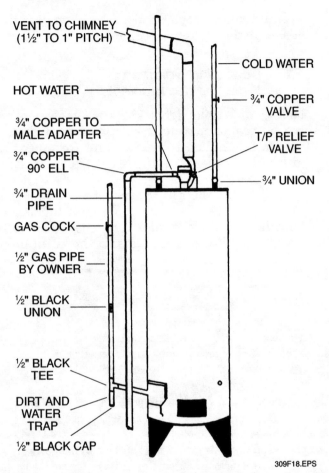

Figure 18 ◆ Typical gas water heater.

Electric water heaters use a submerged element to heat water (see *Figure 19*). Before replacing an electric water heater element, disconnect the power supply, and then drain the water storage tank. When the tank has been drained, remove the element. Inspect the element for any clues as to the cause of the problem. If the element has excessive calcium buildup, install a water softener. Replace the element with a new one when the problem has been corrected. Tighten the new element to ensure a watertight seal, and then refill the tank.

If the thermostat is out of adjustment, find the correct setting. As discussed in the section called *Hard Water Deposits*, thermostats are extremely sensitive, so adjust thermostats in small increments. Make the first adjustment and show the customer

how to make further adjustments. If the hot water pipes are losing heat, insulate the supply lines.

Sometimes there may be not enough or too much hot water. Conduct a survey of hot water needs. Part of a plumber's job is to offer advice to customers on how to assess their water heater requirements over time. For example, if the water heater is in a home, identify the family's past, current, and future usage. The family may have grown children who have recently left home. As a result, they may be using less hot water. If so, the thermostat could be lowered or the water heater could be replaced with a smaller unit. You can also suggest how the rest of the family can reduce their hot water usage. Walk the customer through the proposed changes. Survey the family's needs and then recommend whether they should make adjustments or replace their water heater. The more you inform your customers, the more likely you will be to keep their business.

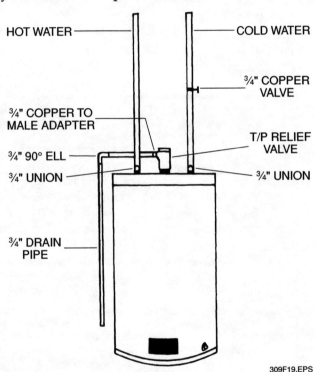

Figure 19 ◆ Typical electric water heater.

3.6.2 Replacing a Water Heater

If the water heater is improperly sized, replace it with one that is adequate for the system. Take the following steps when removing a water heater:

Step 1 Shut off the water supply and drain the tank.

Step 2 Disconnect the heater's inlet and outlet water lines. Mark each line connected to the heater. This will help you identify the correct connection when installing the new water heater.

Water Heater Stands

Water heaters are often located near storage areas for flammable liquids such as gasoline and paint thinner. These liquids emit vapors that are heavier than air. These vapors thus tend to puddle near the floor. If they are drawn into a gas-fired water heater, they can cause an explosion. Manufacturers offer hot water heater stands that raise the heater off the floor. The heater can then draw air from above vapor puddles. Consult the local code about the use of hot water stands in your area.

Step 3 Carefully disconnect the water heater's gas piping and electrical connections. If you do not have experience with handling utility connections, seek help from an experienced plumber.

Step 4 Remove the water heater according to the manufacturer's instructions. Consult with an experienced plumber on the best way to remove a water heater or check the local code for guidance. Be sure to avoid spilling or splashing any remaining water in the tank as you remove the water heater from the building.

Before disposing of an old water heater, inspect the elements (if it is an electric water heater) or the inside base of the tank (if it is gas fired). If there is excessive **corrosion** or calcium deposits, suggest that the water be analyzed. Also suggest to the customer that a water softener be installed because water softeners usually allow a water heater to run more efficiently and last longer.

To install a new water heater, follow these steps:

Step 1 Locate and install the new water heater according to the manufacturer's instructions. Consult the local code.

Step 2 Connect all water, gas, oil, or electric supply lines. The lines should have been marked when the old tank was removed.

 WARNING!

The T/P valve is a critical safety control feature in a water supply system. The valve must always function properly. A malfunctioning T/P valve can cause a water heater to explode, resulting in injuries or death. Install the T/P valve so that it drains as an indirect waste line. Always check to ensure that T/P valves are operating correctly and that they are clear of obstructions.

Step 3 Replace the temperature and pressure relief (T/P) valve with a new one before operating the new water heater.

Step 4 Fill the tank with water. The tank must always be filled before the heating elements or burners are turned on. If this is not done, the elements will burn out.

Step 5 Ensure that all air is bled through the hot water faucets.

Step 6 Set the thermostat to the lowest setting that will serve the structure adequately.

Step 7 Ensure that the relief valve drain does not create a trap. A trap will cause back pressure when steam is discharged through the relief pipe.

Step 8 Provide an air gap at the outlet end of the relief valve pipe. The air gap should be located where the valve empties into the floor drain or other approved receptacle.

3.7.0 Servicing Underground Pipes

Replace underground pipes that have been damaged beyond repair. Underground pipes that are undersized will also need to be replaced. Consult an expert plumber for guidance on how to replace pipe in your area.

In most cases, old buried supply pipe should be abandoned. Plan a new route for the replacement pipe by studying the existing site and structure. Note the location of all nearby utility lines before you start to dig. These include gas, electric, cable, and phone lines. In the structure, seal all old branches back to the old main or service pipe. Ensure that the new supply line connects with all existing branches in the structure.

Review Questions

Sections 1.0.0–3.0.0

1. __C__ is/are the most common problem(s) that plumbers encounter in water supply systems.
   a. Low water pressure
   b. Frozen pipes
   c. Leaking pipes
   d. Poor water quality

2. Use a __b__ to temporarily repair a leak in a plastic or copper pipe.
   a. pipe cutter
   b. pipe clamp
   c. water hammer arrester
   d. electric heat tape

3. Ice in a pipe causes the pipe to __A__.
   a. expand
   b. contract
   c. lengthen
   d. corrode

4. Hard water contains large amounts of __C__.
   a. sodium
   b. charcoal
   c. minerals
   d. metals

5. When replacing a water heater, the plumber should __D__ the T/P valve.
   a. repair
   b. clean
   c. inspect
   d. replace

4.0.0 ◆ SERVICING DWV SYSTEMS

In *Plumbing Level Two*, you learned how to design and install drain, waste, and vent (DWV) systems. DWV systems are vital for public health and sanitation. A well-designed DWV system will usually operate smoothly and efficiently. However, problems can and do crop up. The most common problems are **clogs** and odors. Plumbers should address and correct drainage problems promptly. If a problem goes untreated, it could cause contamination, illness, and even death. The most common procedures for servicing DWV systems are discussed below.

4.1.0 Clogs

Clogs are the most common problem in waste drainage systems. A clog is a pipe blockage that interferes with the flow of waste liquids. Solid waste matter, tree roots, rust, ice, and other debris are common causes of clogs. Clogs may occur in any part of a DWV system. Clogs in drain lines commonly happen at pipe bends, in fixture traps, and in improperly pitched sections of pipe. Vent pipes can also become clogged. In this section, you will learn how to identify and fix different types of clogs in DWV systems.

4.1.1 Drain and Trap Clogs

Fixture traps protect against contamination from waste products. They do this by maintaining a water seal in a U-shaped length of pipe. However, the shape of a trap makes it easy for debris to collect there as well. When too much debris collects in the trap, water may be slow to drain and clogs may cause water to back up. A clog in a fixture trap can usually be reached through the fixture's drain. Clogs in fixture traps are more of a nuisance than a serious problem.

To unclog a sink or lavatory drain, use a plunger (see *Figure 20*). Note that plungers are less effective on vented pipes. This is because the pressure created by the plunger is released through the vent. The same is true for fixtures that have overflows,

> **WARNING!**
> DWV lines may contain toxic and flammable vapors. The public sewer is also a biohazard. Use appropriate personal protective equipment when working with DWV systems. Wear hand and eye protection as necessary.

such as bathroom lavatories and bathtubs. To plunge a fixture with an overflow, first block the overflow with a plug such as a wet rag. This will keep the pressure created by the plunger from escaping.

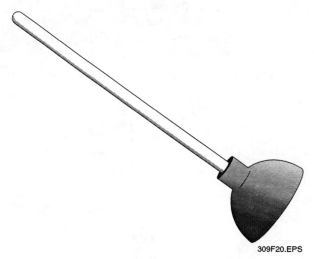

Figure 20 ◆ Plunger.

When plunging does not clear the clog, use a **snake** (see *Figure 21*). A snake, also called an *auger*, is a long, flexible metal wire wound on a drum. On the end of the wire is a head designed to break up a clog. Electric snakes are also available for large or tough clogs. Use the following steps to snake a drain:

Step 1 Remove the drain plug from the fixture.

Step 2 Insert the end of the snake into the drain. Work the snake deeper into the drain by pushing and spinning the handle. This will allow the snake to move through bends and fittings.

Step 3 When the snake contacts the clog, spin the snake handle until it breaks through the clog. Always spin the handle in one direction only. Pushing and pulling the snake back and forth also helps break up the clog.

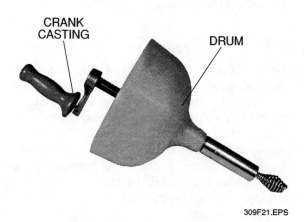

Figure 21 ◆ Hand snake.

Step 4 When it feels like the clog has been cleared, run water through the drain. If the water flows freely, the clog has been cleared. If the water does not flow freely, repeat Step 3.

Step 5 When the clog has been completely cleared, remove the snake and run water through the drain to flush the debris out of the system.

If neither a plunger nor a snake removes the clog, disassemble and thoroughly clean the trap. Remember to place a bucket or other container underneath the assembly when disassembling a trap. Couplings and washers are designed for easy removal (see *Figure 22*). If the washers are stiff or damaged, replace them. Inspect the inside of the trap assembly for rough edges and projections, because these can cause a clog. Some traps have cleanout plugs at their base for easy removal of clogs (see *Figure 23*).

Use a plunger to remove a clog in a toilet. More serious clogs will require a snake. Do not use chemical cleaners to remove toilet clogs. The water will usually dilute cleaners before they can reach the clog. Never flush a clogged toilet until the blockage has been completely broken up and removed. Otherwise, the toilet could overflow.

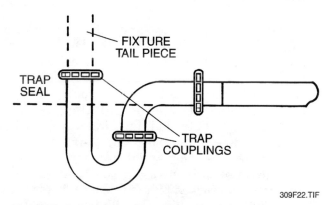

Figure 22 ◆ Location of couplings on a fixture trap.

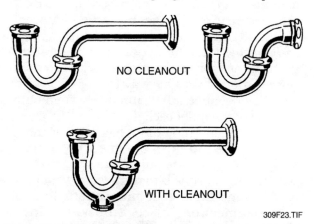

Figure 23 ◆ Examples of fixture traps with and without cleanouts.

After clearing the blockage, test the fixture by flushing several times with toilet paper. If any blockage remains in the line, the paper will catch on it and back up the flow.

In older buildings you may find drum traps (see *Figure 24*). A drum trap is a metal cylinder installed in the fixture waste line. The top of the trap sticks up through the finished floor for easy access. Drum traps are now prohibited by most plumbing codes. Local plumbing authorities must approve the use of a drum trap before it can be installed. Jewelry shops are commonly permitted to use drum traps to trap stone and gem particles. Access to drum traps is usually through the basement or a ceiling access panel.

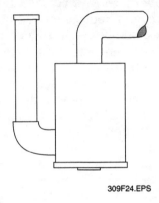

309F24.EPS

Figure 24 ◆ Drum trap.

Tree roots, grease, ice, and other debris often cause clogs in the building drain. DWV systems are designed so that cleanouts are accessible to plumbers, because cleanouts allow plumbers to clean pipes that cannot be easily reached through drains and traps (see *Figure 25*). Use a snake to break up clogs in the building drain. To clear tree roots, use an electric snake with a special flexible cutting blade (see *Figure 26*). Tree roots will continue to cause problems as long as the trees remain and until the cracked pipe is repaired. Replace the drain and dig a new sewer with good joints such as glued PVC or cast iron pipes. Before excavating, be sure to call utilities protection services, and follow all OSHA excavation guidelines as covered in *Plumbing Level Two*. Local horticulturalists can provide information about trees with roots that do not threaten pipes. One way to retard root growth is to flush copper sulfate into the system with the last use of the toilet each day.

When the building drain has been cleaned, use a garden hose to flush the line with water. This will wash the debris out of the system. Another way to clear the line is to run several water outlets at the same time. Take precautions to avoid a cross-connection. Place a bucket below the cleanout to protect the floor from spills.

> **WARNING!**
> Inserting a garden hose into a cleanout sets up a cross-connection between the potable water supply and the waste line. Install a vacuum breaker on the hose bibb if it does not have one. Otherwise, flushing the drain may cause a sudden backflow that will contaminate the fresh water supply.

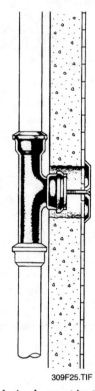

309F25.TIF

Figure 25 ◆ Building drain cleanout installed in a wall.

309F26.EPS

Figure 26 ◆ Root-cutting attachment for a snake.

4.1.2 Vent Clogs

Clogs sometimes occur in vent stacks rather than in drains. Bubbling water in sink and bathtub drains indicates a clogged vent. Vent clogs can be caused by solids backing up into the vent from the drain line or by outside debris falling into the vent. In cold climates, frost and heavy snow can plug a vent opening. Plugged vents prevent the air pressure in a DWV system from equalizing, which can lead to trap siphonage and the escape of sewer odors.

Sewer odors indicate dry trap seals, which can be caused by a restricted vent opening. Clean vent stacks from where they exit the roof. Insert a hand or electric snake through the vent terminal to clear the blockage. When the blockage has been completely cleared, flush the vent using a garden hose or an inside tap. This will keep the debris from clogging the drain line at another location.

If the clog is in a branch vent, you may have to tear out part of a wall to gain access to the vent piping. If the branch line is both a drain and a vent, use an electric snake inserted through the exposed portion of the vent. This procedure may be easier than going through a sink drain if the clog is beyond the branch line.

During the winter, vapors from the DWV system can freeze at the vent opening when they meet cold outside air. As layers of frost continue to build up, the vent opening will become more and more restricted. Eventually, the frost will close the vent. One way to solve this problem is to enlarge the vent opening (see *Figure 27*). Note that the enlarged section of vent stack must extend at least one foot below the roofline to be effective.

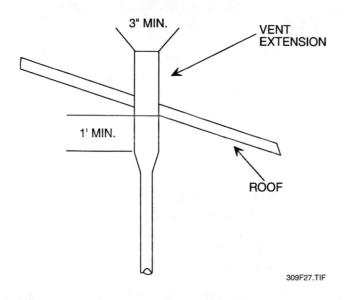

Figure 27 ◆ An enlarged vent opening prevents frost buildup.

Repeated trap siphonage means that the vent is undersized. An undersized vent prevents the air pressure in the DWV system from equalizing. When this is the problem, replace the vent.

4.1.3 Clogs in Restaurants

Many restaurants are designed with separate waste lines. Separate lines for water-closet waste, bar sinks, and kitchen waste prevent contamination of all fixtures from a backup in one. When clearing a clog in a restaurant waste line, make sure you are snaking the correct line. You may need to consult the building's plumbing drawings. Take precautions to protect food-handling fixtures from spills and splashes.

4.2.0 Odors

Sewer odors indicate dry trap seals or leaks in the waste line. Vents that exit the roof near air intakes for air conditioning or ventilating systems can also cause odors to enter a structure. In addition to siphonage, several things can cause the loss of a trap seal. If the fixture has not been used in a long time, the water in the seal may have dried up. Tight door seals or exhaust fans in public restrooms can also cause a trap to lose its seal. Install trap primers to correct dry traps (see *Figure 28*).

Sewer gases can contain hydrogen sulfide, which, in large amounts can destroy copper DWV piping. To prevent the buildup of hydrogen sulfide, some codes permit the installation of a house trap on the building drain. House traps on building drains can be difficult to access. The International Plumbing Code® permits house traps only when required by local code officials. When installing a house trap, provide the trap with a cleanout and install a relief vent or other air intake on the inlet side of the trap. Consult the local code for other requirements related to house traps.

If there is an odor in a DWV system and it has a sewage removal system, check the seating of the sump lid (see *Figure 29*). Check the sump vent for blockage. Some commercial installations such as Laundromats use special lint traps. Inspect the lint trap and clean it if it is the source of the odor.

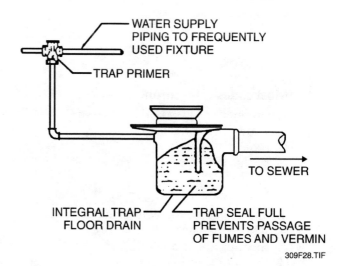

Figure 28 ◆ Trap primer installed in a DWV system.

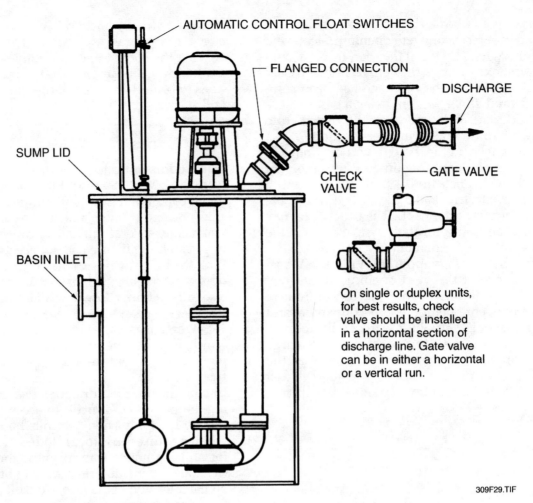

AUTOMATIC CONTROL FLOAT SWITCHES

FLANGED CONNECTION

DISCHARGE

SUMP LID

CHECK VALVE

GATE VALVE

BASIN INLET

On single or duplex units, for best results, check valve should be installed in a horizontal section of discharge line. Gate valve can be in either a horizontal or a vertical run.

309F29.TIF

Figure 29 ◆ A typical sewage removal system.

Review Questions

Section 4.0.0

1. Clogs can be accessed through a fixture drain or a ___.
   a. cleanout
   b. fitting
   c. trap
   d. hose bibb

2. Most codes ___ drum traps.
   a. prefer
   b. allow
   c. require
   d. prohibit

3. To clear a clog in a vent, insert the snake ___.
   a. at the base of the vent stack
   b. where the vent exits the roof
   c. through the branch drain
   d. through the cleanout

4. Sewer odors indicate ___ or dry trap seals.
   a. leaks in the waste line
   b. leaks in the vent
   c. clogged traps
   d. backflow

5. To prevent the buildup of hydrogen sulfide, some codes permit the installation of a(n) ___.
   a. deionizer
   b. activated charcoal filter
   c. house trap
   d. drum trap

5.0.0 ◆ PIPE CORROSION

Exposure to air, water, and soil can cause water supply piping and DWV piping to deteriorate. In plumbing, this is called *corrosion*. Corrosion happens when pipe metal reacts to chemicals in the surrounding environment. Leaky pipes may indicate that corrosion has eaten away at the pipe. If one section of pipe is corroded, you can expect to find other sections of pipe corroded. Inspect the entire system for signs of deterioration.

Rust is a common form of pipe corrosion. Rust is a powdery or flaky residue that forms on iron and steel pipes exposed to moisture and air. If not treated, it can eventually eat through the pipe and cause a leak. If steel pipe is rusting, identify any sources of moisture in contact with the pipe and take steps to correct the problem. For example, if the pipe is in a humid basement, install dehumidifiers. If you find no specific water problems, replace rusted pipe with copper or plastic pipe. Use copper pipe for water supply lines and plastic for waste lines. Follow the manufacturer's specifications when installing new pipe.

Pipes buried in soil can encounter a form of corrosion known as **electrolysis**. Electrolysis breaks down the chemical composition of a pipe through long-term exposure to a small electrical current. When the pipe has a stronger electrical current than the soil, electrolytic breakdown will happen. Replace deteriorating pipe with nonmetallic pipe.

An alternative is to install a sacrificial anode on the line (see *Figure 30*). Sacrificial anodes undergo electrolysis instead of the pipe run, protecting the pipe. Install anodes on steel and wrought iron pipe runs. Codes also require anodes on plastic pipes that use steel and iron couplings. Refer to the local code.

Another common form of pipe deterioration is **galvanic corrosion**. Galvanic corrosion can happen when two different types of metal are joined. Contact between the two metals can actually create a small electrical current. This current causes metal ions to flow from one pipe to the other (see *Figure 31*). This action slowly erodes the metal in the pipe that loses the ions. To stop the ion flow, install dielectric fittings to break the connection between the different pipes.

Soil with a high acid content can also cause galvanic corrosion. Test the soil for acid content and refer to the local code for acid levels in your area. If the soil is the cause of the corrosion, dig up and replace the pipe with pipe that more closely matches the soil. Talk with an experienced plumber to learn how to identify acidic soils.

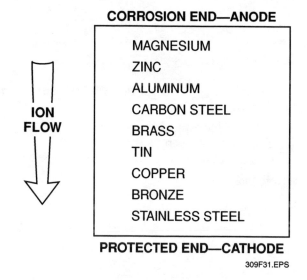

Figure 31 ◆ Galvanic scale.

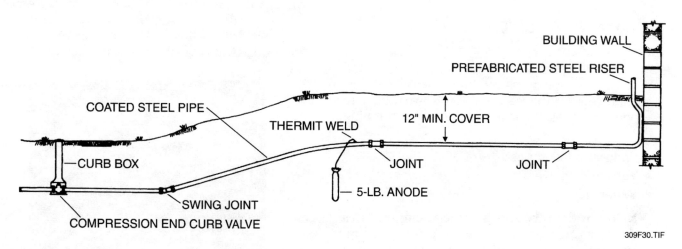

Figure 30 ◆ Sacrificial anode installed on an underground gas line.

Review Questions

Section 5.0.0

1. _____ is not caused by an electrical charge.
   a. Rust
   b. Galvanic corrosion
   c. Electrolysis
   d. Ionization

2. The most common causes of corrosion are exposure to air, water, or __a__.
   a. metals
   b. salt
   c. mineral salts
   d. soil

3. When replacing rusted pipe, use __b__ pipe for the water supply lines.
   a. plastic
   b. copper
   c. stainless steel
   d. dielectric

4. Install a __a__ to protect buried pipes from electrolysis.
   a. sacrificial anode
   b. dielectric fitting
   c. pipe clamp
   d. deionizer

5. Acidic soil can corrode a pipe through exposure to _____.
   a. mineral salts
   b. electrolysis
   c. galvanic corrosion
   d. leaching

6.0.0 ◆ SERVICING PLUMBING FIXTURES

In *Plumbing Level Two*, you learned how to install basic plumbing fixtures and how to service valves and faucets. In this section, you'll learn how to service and replace the following basic plumbing fixtures:

- Sinks and lavatories
- Tubs and showers
- Water closets

Many of the procedures for replacing fixtures and fittings are basically the same as for installing new ones. At this point in your training, you should know what tools and materials are required to repair and replace plumbing fixtures. Ask your instructor if in doubt.

> **CAUTION**
>
> A new fixture or appliance may not have the same dimensions as the one it is replacing. Always check the rough dimensions to verify measurements and placement.

Fixtures can become cracked or chipped during use. Replace damaged fixtures for both aesthetic and safety reasons. Before servicing or replacing fixtures, note the locations of the water supply lines and waste lines. Consider the ease of hookup when selecting a new unit.

6.1.0 Repairing and Replacing Sinks and Lavatories

Leaking faucets are a common problem with sinks and lavatories. To correct a leak in a cartridge faucet, replace the cartridge and seals. Drips in rotating ball faucets usually occur as a result of wear on the valve seals. Replace the valve and springs. Leaks from the base of a rotating ball faucet are caused by worn O-rings. Replace with a properly sized O-ring. Ceramic disc faucets can leak as a result of mineral deposits on the inlet ports. Clean the inlet and replace the seals.

When replacing a damaged sink or lavatory in an existing countertop, discuss the following options with the customer. Select a fixture that will fit the shape and size of the cutout. Ensure that the unit is the right color and material. Determine whether the unit should have a self-sealing rim (see *Figure 32*) or a metal rim (see *Figure 33*). Always test the unit to make sure it operates properly.

Clean the countertop before installing the new unit. For a sink with a self-sealing rim, set the unit directly on a thick bead of silicone caulk. Remove excess caulk from the fixture and countertop before it dries. Allow the silicone caulk to set for 24 hours. That will allow the caulk to cure fully.

309F32.EPS

Figure 32 ◆ Sink with a self-sealing rim.

Figure 33 ◆ Sink with a metal rim.

Sealing metal rim sinks and lavatories with silicone will prevent water from running under the countertop. Use the following steps to install a sink with a metal rim:

Step 1 Press the rim into a silicone bead on the countertop.

Step 2 Run a second silicone bead on the sink edge.

Step 3 Press the sink into the second bead.

Step 4 Tighten the sink mounts to secure the sink to the counter.

Step 5 Attach the drain and water supply assemblies. Follow the manufacturer's installation instructions.

CAUTION

Always read and follow the instructions on the side of primer, solvent, and silicone containers. The manufacturers' safety precautions, handling tips, and drying times are printed there.

6.2.0 Repairing and Replacing Tubs, Showers, and Water Closets

In the past, tubs and showers were designed as one-piece units. Such units were too large to pass through doors. Nowadays, lightweight fiberglass tubs and shower stalls come in sectional units that can be installed easily. This makes it easier to replace damaged units or upgrade to fancier ones. Plumbers repair and replace tubs and showers. Follow the manufacturer's installation directions. Take precautions to avoid damaging the units during installation.

Manufacturers often recommend laying a wet bed of mortar on the floor before installing a fiberglass tub or shower base. Set the base of the new unit into the mortar carefully and make sure the fit is correct. Let the mortar harden according to the mortar instructions. The mortar will conform to the base design as it dries. This gives maximum support to the fiberglass unit (see *Figure 34*).

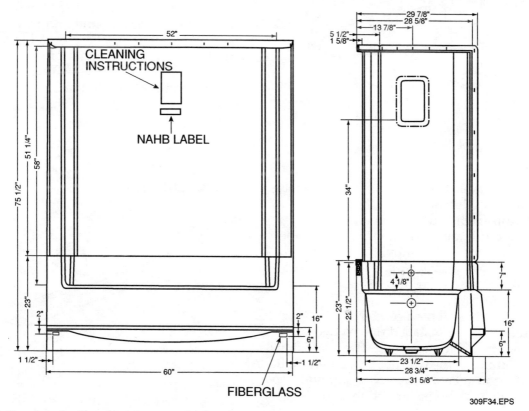

Figure 34 ◆ Rough-in for fiberglass shower and tub unit.

Shower sidewalls require a square enclosure. When installing ceramic tiles on the sidewalls, install water-resistant sheetrock as a backing. Remove the old wall material if necessary. In most cases, install the fiberglass unit directly against the 2 × 4 wall studs. Shim the opening to square it up. Install bracing for the tub (see *Figure 35*).

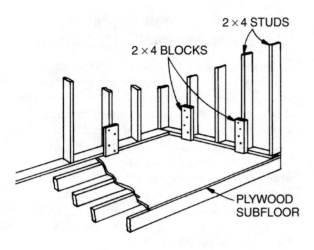

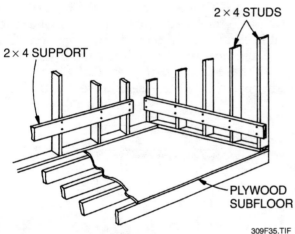

Figure 35 ◆ Bracing for the tub.

The existing piping will determine which side of the shower or tub will have the faucets and drain. There are right- and left-hand units (see *Figure 36*). Some tub and shower units have holes already cut for the water supply outlets. Follow the manufacturer's instructions closely when cutting into fiberglass. Wear appropriate personal protective equipment.

Showerheads in stalls and combination tub and shower units may be installed one of two ways. One option is through a hole in the fiberglass unit. The other is through the wall above the unit. Both locations are considered correct. Consult with the customer to determine the desired location.

309F36.EPS

Figure 36 ◆ Left-hand shower stall unit.

Customers may want to replace water closets for several reasons. The bowl or tank may be cracked or damaged, or the customer might want a different style of fixture. If a bowl or tank is leaking, disassemble the unit and inspect the gaskets and seals. If they are worn, hard, or broken, replace them. One common source of leaks in water closets is the wax seal between the floor and the base of the unit. Replace a worn or damaged seal with a new one (see *Figure 37*).

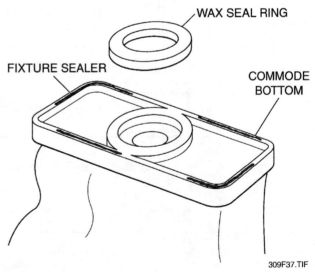

309F37.TIF

Figure 37 ◆ Location of the wax seal.

6.2.1 Troubleshooting Flushometers

Flushometers (*Figure 38*) are installed on commercial water closets. The troubleshooting guide in *Table 1* applies to flushometers. Four common problems are associated with flushometers:

- Leakage around the handle
- Failure of the vacuum breaker
- Malfunction of the control stop
- Leakage in the diaphragm that separates the upper and lower chambers

Kits that contain the components to repair each of these defects are available. Even though all flushometers contain the same basic parts, specific parts vary from manufacturer to manufacturer. Always follow the manufacturer's specifications and instructions when installing replacement parts.

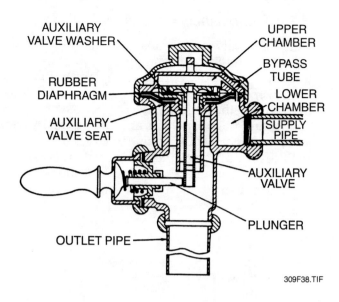

309F38.TIF

Figure 38 ◆ Diaphragm-type flushometer.

Table 1 Troubleshooting Guide for Flushometers

Problem	Cause	Solution
1. Nonfunctioning valve	Control stop or main valve closed	Open control stop or main valve
2. Not enough water	Control stop not open enough	Adjust control stop to siphon fixture
	Urinal valve parts installed in closet parts	Replace with proper valve
	Inadequate volume or pressure	Increase pressure at supply
3. Valve closes off	Ruptured or damaged diaphragm	Replace parts immediately
4. Short flushing	Diaphragm assembly and guide not hand-tight	Tighten
5. Long flushing	Relief valve not seating	Disassemble parts and clean
6. Water splashes	Too much water is coming out of the faucet	Throttle down the control stop
7. Noisy flush	Control stop needs adjustment	Adjust control stop
	Valve may not contain quiet feature	Install parts from kit
	The water closet may be the problem	Place cardboard under toilet seat to separate bowl noise from valve noise— if noisy, replace water closet
8. Leaking at handle	Worn packing	Replace assembly
	Handle gasket may be missing	Replace
	Dried-out seal	Replace

6.2.2 Troubleshooting Tank Flush Valves

Tank flush valves are available in many styles (see *Figure 39*). If the valve and the lever that operates the valve are badly corroded, replace both parts. However, most problems occur with the component parts—the tank ball, the flapper tank ball, the connecting wires or chain, and the guide. Inspect all of these parts to determine whether to repair or replace the entire assembly. If the valve seat is corroded, you can use a reseating tool to restore it. Always consult with the customer before replacing a fitting.

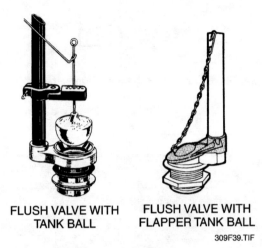

FLUSH VALVE WITH
TANK BALL

FLUSH VALVE WITH
FLAPPER TANK BALL

309F39.TIF

Figure 39 ◆ Tank flush valves.

CAUTION
Never try to adjust the water level in a water closet by bending the float arm of a float-controlled valve. If the rod works around a half-turn (180 degrees), this will cause the tank to overflow.

DID YOU KNOW?
Jacuzzi Whirlpool Bath

In 1900, the Jacuzzi brothers came to the United States from Italy. They were an inventive family. They designed and built an airplane that carried airmail for the post office. They also invented and marketed deep-well pumps. In the late 1950s, the brothers' company turned its attention to developing a portable whirlpool pump for bathtubs. In 1968, Roy Jacuzzi invented the first self-contained whirlpool bath. Two years later, the company developed its first spa model. Today, the company has manufacturing plants around the world, and the name *Jacuzzi* is synonymous with whirlpool baths.

6.2.3 Troubleshooting Ball Cocks

Ball cocks are float-controlled valves (see *Figure 40*). They are installed in water closet flush tanks. Ball cocks control flow by maintaining a constant water level in the tank. Ball cocks can wear out with repeated use. Replace a worn or damaged ball cock rather than repairing it. To install a new ball-cock valve assembly, follow these steps:

Step 1 Turn off the water supply.

Step 2 Lower the threaded base through the bottom hole in the closet tank with the gasket in place against the flange.

Step 3 Place the washer and nut on the base of the assembly on the outside of the tank and tighten. Do not overtighten the nut.

Step 4 Place the riser and coupling nut on the base of the float valve and hand tighten.

Step 5 Align the riser with the shutoff valve. Remove the riser and cut to fit.

Step 6 Reassemble the riser to the tank and valve, and then turn on the water supply.

Step 7 Adjust the float in the tank to achieve the water level indicated on the inside of the tank.

Step 8 Adjust the float with the adjusting screw. Do not bend the float arm to adjust the water level.

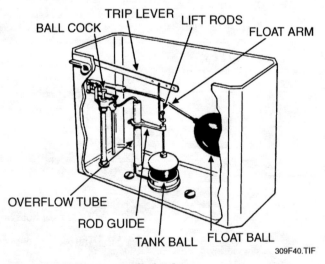

TRIP LEVER   LIFT RODS
BALL COCK   FLOAT ARM

OVERFLOW TUBE
ROD GUIDE
TANK BALL   FLOAT BALL

309F40.TIF

Figure 40 ◆ Float-controlled valve.

CAUTION
Never overtighten the nut on the base of the assembly on the outside of the tank. Overtightening puts stress on the fixture and can cause it to crack.

7.0.0 ◆ INSTALLING ADDITIONAL FIXTURES AND APPLIANCES

You have learned that plumbers design, install, and repair systems and fixtures. Plumbers are also called in to work on remodeling and renovation projects. House additions need fixtures and appliances, too. In most cases, it makes sense to tie the new units into existing water supply and DWV systems. Plumbers ensure that the new lines are correctly designed and adequately sized. Plumbers are also responsible for calculating the water needs of the modified system.

7.1.0 Domestic Dishwashers

When installing dishwashers (see *Figure 41*), consider the following factors:

- Available space
- Access to the hot water supply
- Location of the waste line

Determine the usage requirements early to help the customer select the proper size appliance. The size of the appliance can affect the location of kitchen cabinets and countertops. The most common sizes of dishwashers are 18-inch and 24-inch units.

Dishwashers use hot water only. Usually, the dishwasher is located next to the sink. This arrangement allows easy access to the hot water line. Install a shutoff valve on the hot water inlet line to allow you to service the dishwasher without having to shut off the water system. Dishwashers need hotter water than other fixtures, usually between 140°F and 150°F, but most water heater thermostats are set lower than this. To solve this problem, install a booster on the dishwasher's hot water inlet line. A booster increases the water temperature for a single appliance.

309F41.EPS

Figure 41 ◆ Domestic dishwasher.

> **CAUTION**
>
> New fittings can have hidden defects. Be sure to examine new fittings for holes, cracks, or poor casting. Read and follow the manufacturer's installation instructions carefully.

Install the dishwasher's waste line so that it empties into the sink drain. If the sink has a garbage disposal, connect the drain hose to the disposal (see *Figure 42*). Check the local code before doing this. Many codes state that the drain hose should first be looped above the fixture's flood-level rim (see *Figure 43*) to reduce the chance of a backup into the dishwasher if the waste line clogs. Some codes require a vacuum breaker or an air gap assembly in the dishwasher discharge piping (see *Figure 44*), which is another way of preventing waste from backing up into the dishwasher.

309F42.EPS

Figure 42 ◆ Domestic garbage disposal.

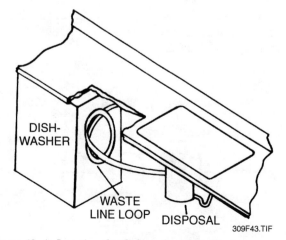

DISH-WASHER

WASTE LINE LOOP   DISPOSAL

309F43.TIF

Figure 43 ◆ Looping the dishwasher drain hose.

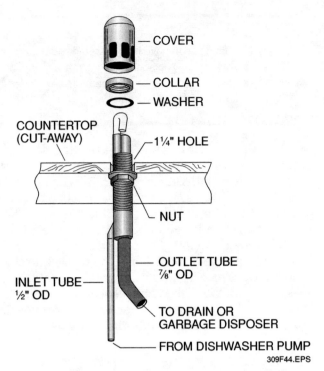

Figure 44 ◆ Air gap assembly for a domestic dishwasher.

7.2.0 Lawn Sprinklers

Plumbers do not always install lawn sprinkler systems. However, they are responsible for hooking them up to the potable water supply. Install a dual-check (DC) valve backflow preventer on the water supply connection (see *Figure 45*). DC valves protect against both back siphonage and back pressure. Install a shutoff valve on the sprinkler supply line to allow access for servicing.

Install sprinkler lines so that they are pitched in the direction of the home. Another option to protect the sprinkler line is to install an automatic drain fitting connected to an underground **dry well**. A dry well is simply a covered hole filled with gravel into which water can drain. Some systems may use air pressure to blow out the lines. Refer to the local code for guidance. Whatever method you choose, remember to drain the water lines before the winter season to prevent damage caused by freezing.

1. Dishwasher waste lines should drain into the _____.
   a. indirect waste line
   b. building drain
   c. sink drain
   d. nearest fixture drain

2. Two ways to prevent backflow in a dishwasher are to install an air gap and to _____.
   a. direct the drain hose into an indirect waste system
   b. direct the drain hose into the hot water inlet
   c. loop the drain hose above the fixture's flood-level rim
   d. loop the drain hose above the building drain

3. If a kitchen sink has a garbage disposal, connect the dishwasher drain hose _____.
   a. the same way as you would if there were no garbage disposal
   b. to the washing machine drain
   c. to the refrigerator drain
   d. to the garbage disposal

4. To protect a lawn sprinkler system from an accidental cross-connection, install a _____.
   a. dual-check valve
   b. double-check valve
   c. reduced pressure zone principle backflow preventer
   d. vacuum breaker

5. Sprinkler lines should be pitched _____.
   a. toward the building drain
   b. no more than one pipe diameter in slope
   c. toward the home
   d. away from the home

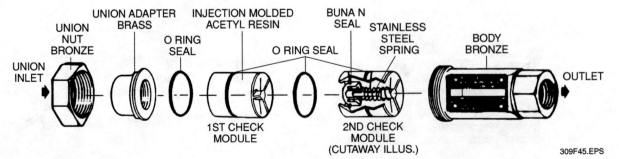

Figure 45 ◆ Dual-check valve assembly.

Summary

When a water supply system, DWV system, fixture, or appliance breaks down, you'll be called in to service it. When customers add new fixtures or appliances, they will call you. Every service call presents new challenges. It is up to you to do the work as professionally and efficiently as possible. To service plumbing fixtures, you need a great deal of knowledge and experience. You are already developing the skills you will use in the field. Service calls are an excellent chance to gain valuable experience. Work with an experienced plumber and ask a lot of questions.

Common problems with water supply systems include leaks, frozen pipes, high or low water pressure, water quality, and flow rate. Problems with water heaters should be corrected promptly. Sometimes the water heater must be replaced.

Clogs in drainpipes can cause problems with DWV systems. These clogs can be cleaned out through fixture drains or cleanouts. Odors can indicate dry traps or leaking pipe. Forms of corrosion, such as rust, electrolysis, and galvanic corrosion, can cause pipes to deteriorate. Plumbers must replace corroded pipe.

Fixtures and appliances also break down. Plumbers repair or replace components to fix the problem. Sometimes damaged fixtures must be completely replaced. Plumbers also install new fixtures as part of renovations or additions. No matter what the service call, plumbers must always arrive promptly and work safely and efficiently. Your professional reputation depends on making sure customers get what they want and need.

Trade Terms Introduced
in This Module

Activated carbon filter: A *water filter* that uses charcoal to absorb sulfur from potable water.

Clog: A blockage in a pipe.

Condensation: Water that collects on a cold surface that has been exposed to warmer, humid air.

Corrosion: Deterioration caused by chemical reaction.

Deionizer: A *water filter* installed on a cold water inlet that filters out metals and mineral salts.

Dry well: A drain consisting of a covered hole in the ground filled with gravel.

Dynamic water pressure: The pressure water exerts on a piping system when the water is flowing (circulating water). Dynamic water pressure, measured in psi, will always be lower than *static water pressure*.

Electric heat tape: An insulated copper wire wrapped around a pipe to keep it from freezing.

Electrolysis: A form of *corrosion* caused by exposure to a small electrical current.

Galvanic corrosion: A form of *corrosion* resulting from the contact of two different types of metal.

Hard water: Potable water that contains large amounts of mineral salts such as calcium and magnesium.

Reverse osmosis method: A method of filtering water through cellulose acetate membranes used in some *water filters*.

Rust: Residue that forms on certain metals exposed to moisture and air.

Schrader valve: A tire valve stem installed in a tank, which is used to correct low pressure.

Snake: A flexible metal wire inserted into a pipe to break up a *clog*.

Static water pressure: The pressure water exerts on a piping system when there is no water flow (noncirculating water). Static water pressure, measured in psi, will always be higher than *dynamic water pressure*.

Wall hydrant: A hydrant installed in a building's exterior wall, which works by draining water after the valve is closed.

Water filter: A device designed to remove small particles and organisms from the potable water supply.

Water softener: A device that chemically removes mineral salts from *hard water*.

Yard hydrant: A lawn spigot connected to the water main, which works by having its valve located below the frost line.

Zeolite system: A *water softener* that filters mineral salts out of *hard water* by passing the water through resin saturated with sodium.

Additional Resources

This module is intended to present thorough resources for task training. The following reference works are suggested for further study. These are optional materials for continued education rather than for task training.

Kitchen & Bathroom Plumbing, How to Fix It Series, Vol. 1, No. 20, 1998. Ron Hazelton, ed., New York, NY: Time-Life Books.

NAPHCC Repair Remodeling Manual, 1984. Falls Church, VA: Plumbing-Heating-Cooling Contractors—National Association.

Plumbing: A Guide to Repairs and Improvements, 1998. Jeff Beneke. New York, NY: Sterling Publishing Company.

References

1997 International Plumbing Code, 1998. Falls Church, VA: International Code Council.

"BOCA Research Report No. 96-69," http://www.boca-es.com/pdf/96-69.pdf, 1998. BOCA Evaluation Services, Inc., www.boca-es.com, December 2001.

Efficient Building Design Series, Volume III, Water and Plumbing, 2000. Ifte Choudhury and J. Trost. Upper Saddle River, NJ: Prentice Hall.

Guide to Plumbing, 1980. New York, NY: McGraw-Hill Book Company.

"History of Jacuzzi," http://www.jacuzzi.com/history, Jacuzzi Whirlpool Bath Inc., www.jacuzzi.com, August 2001.

Planning Drain, Waste & Vent Systems, 1993. Howard C. Massey. Carlsbad, CA: Craftsman Book Company.

"The Need for Qualified Medical Gas Piping Installers, Inspectors, and Verifiers", *Plumbing Standards Magazine* (January - March), 2000. Dale Dumbleton. Westlake, OH: American Society of Sanitary Engineering.

"What it Takes to be a Certified Medical Gas Installer, Inspector, and Verifier", *Plumbing Standards Magazine* (January - March), 2000. Ron Ridenour. Westlake, OH: American Society of Sanitary Engineering.

Figure Credits

Crain Company	309F39
Culligan International	309F13, 309F15
Fluidmaster	309F40
In-Sink-Erator	309F42
Kohler Company	309F29, 309F32, 309F33, 309F36
Mansfield Plumbing Products	309F38
Marco Products Company	309F21, 309F26
Maytag Corporation	309F41
Owens/Corning Fiberglass	309F34
Sioux Chief	309F06
The Spudgun Technology Center/Joel Suprise	309F11
Watts Regulator Company	309F08, 309F45

NCCER CRAFT TRAINING USER UPDATES

The NCCER makes every effort to keep these textbooks up-to-date and free of technical errors. We appreciate your help in this process. If you have an idea for improving this textbook, or if you find an error, a typographical mistake, or an inaccuracy in the NCCER's Craft Training textbooks, please write us, using this form or a photocopy. Be sure to include the exact module number, page number, a detailed description, and the correction, if applicable. Your input will be brought to the attention of the Technical Review Committee. Thank you for your assistance.

Instructors – If you found that additional materials were necessary in order to teach this module effectively, please let us know so that we may include them in the Equipment and Materials list in the Instructor's Guide.

Write:	Curriculum Revision and Development Department
	National Center for Construction Education and Research
	P.O. Box 141104, Gainesville, FL 32614-1104
Fax:	352-334-0932
E-mail:	curriculum@nccer.org

Craft _____ Module Name _____

Copyright Date _____ Module Number _____ Page Number(s) _____

Description _____

_____

_____

_____

(Optional) Correction _____

_____

_____

(Optional) Your Name and Address _____

_____

_____

Plumbing Level Three

Index

Index

Indirect waste systems, 4.1–4, 4.13
  defined, 4.14
  piping installation, 4.2–4
Indirect water heaters, 8.17
Individual vents, 3.8, 3.15
Instantaneous water heaters, 8.17
Interceptors, 4.5–11, 4.13
  baffles, 4.5, 4.14
  climb-in interceptors, 4.8
  defined, 4.14
  draw-off hoses, 4.6, 4.9, 4.14
  grease interceptors, 4.3, 4.6–8
  installation of, 4.5–6
  oil interceptors, 4.8–10
  sediment interceptors, 4.10–11
Intermediate atmospheric vent vacuum breaker, 7.9, 7.14
Intermittent demand, 8.7, 8.8, 8.22
International Association of Plumbing and Mechanical
    Officials. *See* IAPMO
International Code Council. *See* ICC
International Plumbing Code®. *See* IPC
Internet resources, 3.8
IPC (International Plumbing Code®), 1.3, 2.15, 3.10, 3.11, 5.3,
    6.7
Isometric drawing, 6.11
Isosceles triangle
  area of, 1.7
  defined, 1.25

Jacuzzi whirlpool bath, 9.26

Kelvin scale, 1.14, 1.15, 1.25
Kilo- (prefix), 1.3
Kilogram (unit), 1.27
Kilometer (unit), 1.27
Kitchen islands, air admittance vents, 3.11
Kitchen sinks
  drainage fixture units of, 5.7
  load values for, 6.8

Laboratories, backflow preventers for, 7.10
Laminar flow, 6.3, 6.16
Laundromats, lint traps, 5.15
Laundry interceptor, 4.10, 4.11
Laundry tubs, backflow preventers for, 7.6
Lavatories
  demand and flow pressure values for, 6.6
  drainage fixture units of, 5.7
  load values for, 6.8
  servicing, 9.22–23
Lead content, of pipes, 8.9
Leaks
  in pressure tank, 8.10
  servicing, 9.3–4
  testing for, 1.19
Legionnaire's disease, health problems caused by backflow
    and cross connection, 7.2, 7.8
Legislation, Safe Drinking Water Act, 9.11
Length, 1.2, 1.3, 1.4, 1.27
Levers, 1.21, 1.25
Lift stations, 5.1, 5.17
Light intensity, units of measurement, 1.3
Limestone, as neutralizing agent, 4.12
Lint traps, 5.15
Liquid measure
  conversion tables, 1.27
  English system, 1.2, 1.10

Liquid thermometers, 1.13, 1.25
Liter (unit), 1.10, 1.25, 1.27
Local codes, saddle fittings, 2.12
Loop vents, 3.9, 3.11, 3.15
Low-flush toilets, 2.8
Low water pressure, 9.6

Machines, 1.20–22
Main vents, 3.4
Maintenance, recirculation systems, 8.20
Manual draw-off systems, 4.9, 4.14
Manufactured air gap, 7.9, 7.14
Mass
  conversion tables, 1.27
  units of measurement, 1.2, 1.3
Math. *See* Applied mathematics
Maximum flow, 8.8, 8.22
Maximum probable flow, 6.10, 8.22
Measurement. *See* Weights and measures
Mechanical advantage, 1.24
Mechanical plug, 1.19
Medical facilities, indirect waste lines, 4.3
Medical gas, 3.3, 9.13
Mega- (prefix), 1.3
Megagram (unit), 1.27
Mercury float switch, 5.6, 5.8, 5.17
Mercury thermometers, 1.13
Meter, 1.27
Methane, 3.6
Metric system, 1.3, 1.4
  conversion tables, 1.27
  defined, 1.25
Metrology, 1.2
Micro- (prefix), 1.3
Microwave oven, 1.13
Mile (unit), 1.2, 1.27
Milli- (prefix), 1.3
Milliliter (unit), 1.27
Millimeter (unit), 1.27
Mineral salts, hard water deposits, 9.8–9
Mineral tank, 9.9
Mini tube cutter, 9.3
Model codes
  adoption of, 2.12–13
  ANSI Safety Requirements for Plumbing, 2.6, 2.7, 2.15
  BOCA National Plumbing Code®, 2.3, 2.15
  change proposal form, 2.10, 2.11
  code standards, 2.10, 2.16–17
  cycle, 2.8, 2.9
  defined, 2.15
  energy conservation and, 2.13
  fittings, 2.12
  IAPMO Uniform Plumbing Code™, 2.5, 2.6, 2.15, 3.10
  ICC International Plumbing Code®, 2.4, 2.5, 2.15, 3.10,
      3.11, 5.3, 6.7
Model Energy Code, 2.13
  new materials, 2.13
  PHCC National Standard Plumbing Code, 2.5–6, 2.7, 2.15
  professional organizations promoting, 2.10, 2.16–17
  review process, 2.12
  revisions, 2.8–12
  SBCCI Standard Plumbing Code™, 2.4, 2.15
  typical chapters in, 2.13
  Uniform Solar Energy Code, 2.13
Model Energy Code (ICC), 2.13
Modified waste line, 4.2, 4.14
Momentum siphonage, 3.2, 3.15